Environmental Policymaking in Congress

ENVIRONMENTAL POLICYMAKING IN CONGRESS
THE ROLE OF ISSUE DEFINITIONS IN WETLANDS, GREAT LAKES AND WILDLIFE POLICIES

KELLY TZOUMIS
Roosevelt University

ROUTLEDGE
A MEMBER OF THE TAYLOR & FRANCIS GROUP
NEW YORK & LONDON
2001

Published in 2001 by
Routledge
A member of the Taylor & Francis Group
29 West 35th Street
New York, NY 10001

10 9 8 7 6 5 4 3 2 1

Library of Congress Cataloging-in-Publication Data

Tzoumis, Kelly.
    Environmental policymaking in Congress : the role of issue definitions in
wetlands, Great Lakes, and wildlife policies / Kelly Tzoumis.
        p.   cm. — (Politics and policy in American institutions ; v. 3)
    Includes bibliographical references and index.
    ISBN 0-8153-3646-2
        1. Environmental policy—United States.   2. Legislative hearings—United
States.   I. Title.   II. Garland reference library of social science. Politics and policy
in American institutions ; v. 3.

GE180.T96 2001
363.7'056'0973—dc21   00-064612

Printed on acid-free, 250-year-life paper
Manufactured in the United States of America

This book is dedicated to my daughter Kaily Walker.
She is my miracle and primary contribution to the future.
I wish for her a better understanding and respect for the earth
than currently available today.

Contents

Foreword

Kelly Tzoumis has written a book that gives insights into policymaking. While those of us who have served in the House and the Senate will not recognize the various models that are described in this book, it should not prevent readers from going through this product to find real insights into how we are dealing with what is a very key issue, the preservation and development of our water resources.

The United States is rich in fresh water resources, the Great Lakes being a prime example. We are roughly 4 percent of the world's population and have about 8 percent of the world's fresh water. Canada is even more fortunate. What we do know is that wetlands are related to replenishing aquifers and that all of these issues referred to in this volume are interrelated. What we need are more and more policymakers and more and more citizens who are willing to look at the long-term implications of our policy in relation to the world's water resources.

This book provides valuable insights.

Senator Paul Simon

Series Editor Preface

The Garland Series, "Politics and Policy in American Institutions" strives to show the interaction of American political institutions within the context of public policymaking. A public policy approach often by definition is all-encompassing. Admittedly, my own interests focus on national policymaking, but the series will also include works on all levels of government. Indeed, I do not want my own specialties to define the series. Therefore, we seek solid scholarship incorporating a wide range of actors, including those outside the usual definition of government actors. The policy concerns, too, are potentially quite broad, with special interests in the policy process and such substantive issue areas as foreign and defense policy, economic and budget policy, health care, social welfare, racial politics, and the environment. The series will publish a considerable range of works, from upper division texts to scholarly monographs, including both hard and soft cover editions.

In this unique sixth volume in the series, Kelly Tzoumis' book fits exactly the series theme of the role of American institutions in public policymaking. Like Hays' *Who Speaks for the Poor*? she incorporates congressional hearings data. This is a fascinating study of environmental policymaking in Congress over its more than 200 year history. It effectively incorporates three models of issue definition each of which reflect three different natural resources (wetlands, great lakes, and wildlife) policies. Tzoumis provides a nice comparison of these three environmental policy sub-issues. *Environmental Policymaking in Congress* is one of the first major works using the nation's entire history of congressional hearings and, thus, one of the most longitudinal studies ever on legislative decision making. As such, it contains fascinating historical material as well as solid empirical analysis.

Tzoumis uses a model of agenda setting based heavily of the important work of Baumgartner and Jones. She considers significant scholarly literature and important variables in her multivariate analysis in which the author develops her own model of issue definition as agenda setting consisting of three elements: dominance, bounded, and valence issues. Tzoumis traces the long historical developments in these three areas of environmental policymaking, finding that wetlands best fits "dominance," Great Lakes is more "bounded," and wildlife is most appropriately labeled the "valence" model. Tzoumis effectively shows the utility of using these issue definition models for understanding agenda setting in Congress. As such, her study of congressional hearings, including types of hearings and witness information, breaks new ground in our understanding of legislative policymaking in an important issue area of public policy. This book is must read for those interested in Congress and/or environmental policymaking.

Steven A. Shull

Acknowledgments

This book is a combination of a five-year process of data collection, research, and analysis. Several students who include Linda Berkowitz, Cindy Bucci, Billie Jean Van Cleave, Kathleen Delaney, Nancy Donati, Linda Finegold, Sue McFaul, David Rubin, and Yoriko Yamao, were instrumental in the successful completion of this research. These people spent countless hours assisting me with data entry, research into historical documents, and helping me manage every aspect of the data set. Many of these students volunteered time beyond the maximum funded by the university. Most of the work landed on Yoriko Yamao who was the final graduate research assistant in the book production process. She primarily helped with the 1995–1999 data. Linda Berkowitz volunteered time with data entry on endangered species amid the many other time commitments she had to balance. Linda Finegold contributed in the early data management of the Great Lakes sections. She has been a fellow investigator and a constant source of research expertise. Kathleen Delaney provided outstanding assistance in editing and making suggestions to better the manuscript. Cindy Bucci, Billie Jean Van Cleave and David Rubin assisted early in the process, helping to initiate the project. Nancy Donati and Sue McFaul also were helpful with data entry and assistance in acquiring information.

Colleagues who provided me intellectual guidance, endured my requests for reviews of the manuscript, and supported me throughout the process include David Hamilton, Dan Krause, Leonard Robins, Margie Rung and Jeff Talbert. Years ago, through the State of Illinois

Governor's office, David Hamilton actually acquired access for me to the congressional hearings. I am much indebted to him. Dan Krause and Margie Rung reviewed chapters and provided helpful comments. Jeff Talbert was always helpful with suggestions and answers to questions. Most of all, besides being a friend, Leonard Robins provided essential comments on the theoretical components of the manuscript which were much appreciated and improved the final product.

Two important people who have always mentored me throughout my education and careers in government and academia are Bryan Jones and Frank Baumgartner. These professors, whenever called upon provided support. Even when others were skeptical, they believed a wide-eyed, young ABD student when she told them she would finish as she left for a job in Idaho. They had faith in my ability and skills. In addition, their advice and role in my life have impacted me in ways of which they are often unaware. By example, they have taught me the importance of quality research, the pursuit of knowledge, and the balance of life with work. I appreciate and admire them in innumerable ways.

I must acknowledge the institutional support of several people at Roosevelt University, including Ronald Tallman, Paul Green, Vinton Thompson, and Stu Fagan, who extended a research leave for the final completion of the writing process. This leave and other institutional support were essential to the final production of the manuscript. In addition, Vicky McKinley has been especially helpful as a co-instructor and colleague while writing the book at Roosevelt University.

Friends who supported me in this process and pushed me forward include Nelda Barnes, Brian Paulson, Janet Karsten, Carol Gebka, John Munro, Fran Smith, Janice Tsuji, and Evie, Patrick, Abi, Zoe, and Isabelle Zuroske. Special mention has to be made of Nelda Barnes, Fran Smith, and Evie Zuroske who are constant sources of strength for me.

Many of the insights to Congressional hearings came from my experience as a Congressional fellow for former Senator Paul Simon of Illinois. These insights, as well as my experience for five years working with the federal government in environmental areas, provided a practitioner's angle to the book. One of my early government managers, Karen Poore, taught me how to implement policy in a large bureaucracy; Jay Hunze, another government supervisor, gave me the freedom and responsibility to grow as a program manager; and Vicky Otten, Bob Shirell, and Sara Spear Smith were important to my experience on Capitol Hill.

I want to acknowledge my family who gave me the time away to work on the book. My husband, David Walker, and grandmother who assisted me with managing the data as well as provided the necessary means for me to spend working on the book. Early in the project, David spent many hours helping with the computer programming elements of managing such a large data set. He often provided computer technical assistance that saved me months of work. Lillian Rozewski, my grandmother, assisted on the wetlands data as well as helped with the household.

Finally, and most of all, I need to acknowledge my daughter Kaily Walker. She is too little to understand her contribution to the book, but she may read this when she gets older and might remember the days in the library or at my office working the computer. She was my overarching inspiration to understanding congressional policymaking and its impact on our environment. I hope I instill in her the respect and love for the environment that I have enjoyed. Like other children, she represents the future. Hopefully, this book will aid policymakers and participants in agenda setting on issues regarding the environment to help protect the planet for future generations.

ENVIRONMENTAL POLICYMAKING IN CONGRESS

CHAPTER 1

The Issue Definition Process in Agenda Setting

When Congress passes a statute, it reflects the outcome of an agenda setting process that typically may include many months, if not years, about discussion of an issue. The nature of that discussion can involve a range of participants and issues from Congressional committees and subcommittees, interest groups or other governmental and nongovernmental parties. Institutional rules and structures add to the parameters in which an issue is considered. Among the institutional factors are Congressional rules for debate, existing agency regulations, and other structures already in place that have a strong influence on the definition of an issue. In public policy, Congress is one of the most important institutional factors impacting the policy debate. Not only does it control the governmental agenda, it determines the level of implementation based on it appropriation powers. Therefore, Congress plays a unique and key role in agenda setting. The political process involved in agenda setting that considers an issue is greatly determined, however, by the mix of participants and the values they introduce into the policy discussion. Knowing who participates in Congressional deliberations of a policy is important for understanding how an issue is defined because this can directly impact implementation of public policy. For instance, the conscious selection of language, use of symbols, reliance on technology or scientific information and experts, can all contribute to the dynamics of how an issue is defined. Thus, one purpose central to this book is to understand how participants in the Congressional process have defined an issue over time. The second purpose that is thread throughout the research is more implicit. It highlights how issue definition impacts public policymaking in Congress and the implications it has on policy implementation.

Rochefort and Cobb (1994) point out that a basic concept for the study of issue definition is ownership: how an issue is characterized helps determine what institutional structures are considered legitimate for addressing it. This recognized authority is perceived as having credibility for identifying causes, consequences, and policy solutions. From an institutional perspective, ownership really connotes jurisdictional control over policy decisions for a particular issue (Rochefort and Cobb 1994). This jurisdictional control is delineated by the committee structure in Congress. In addition, ownership is extended outside of Congressional committees to include both nongovernmental participants and governmental agencies that have a perceived legitimate position in the issue debate.

The study of issues and how they are defined has flourished in the last several years (Rogers and Dearing 1988; Rogers, Dearing, and Bregman 1993). One reason for this interest reflects how the politics of issue definition impacts the success or failure of a policy. Many scholars have examined the role media plays in conveying the tone in agenda setting for the public while not addressing how Congressional participants initially define issues. There is a renewed interest in policy studies that focuses on how society defines public issues. This book contributes to that collection of information by integrating the study of issues with our understanding of agenda setting and Congress.

The Congressional literature shows that elected officials have to be concerned with re-election (Fiorina 1989; Mayhew 1974; Scher 1960) as a self-interest. It has also been demonstrated that this is not the only interest that is a concern of the elected official (Fenno 1978; Hinckley 1975). Congressional members are concerned about jurisdiction of their committees (King 1994; 1997; Price 1978) and having their interests reflected in public policies. Obviously, different Congressional members will have different definitions and views of a particular issue that is considered (Hall 1997). Discussion of an issue can include a range of interests from an extremely narrow scope to one of different, and sometimes, competing interests (Schattschneider 1960). This is particularly true of environmental policies that can involve a large span of interests, which sometimes have intense conflicts, such as those exhibited among industry, agriculture, recreation, public health, preservation, or even international relations. In most policy areas, Congressional participants help define an issue based on what they perceive as a benefit to their interests. How this definition process can be identified and tracked over the life span of a public policy is the subject of this book. A policy reflects how an issue is understood by policymakers at one point in time, and how changes in that understanding, if any, take place over time. Clearly, how issues in Congress become defined, and even redefined over time, plays a meaningful role in the public policy that is implemented.

There is vast literature on who participates in policymaking; some focused on elites (Dye 1999) and others on pluralistic groups (Dahl 1961; Truman 1951). Researchers have generally attempted to understand who participates without linking the tone or perspective each participant brings to the definition of the issue. Certainly, who participates in public policy is an important question, however, this investigation goes further. This study not only asks who participates, but also investigates how the participant has defined the issue and asks what tone the participant brings. The tone that comes from the definition of the issue is the link missing in the literature on agenda setting.

Issue Definition in Agenda Setting

In general, before laws are enacted, the underlying policy is discussed in hearings conducted by Congressional committees. Often a policy reaches Congressional attention because it has been framed in a manner that requires governmental action. Cobb and Elder (1983) describe how scope, intensity and visibility of an issue can help shape how that issue is debated and perceived by policymakers. If the issue is framed in a manner in which it can be addressed by government, then it will have a better chance of capturing Congressional attention. Without this initial element of framing the issue, the governmental agenda is not activated. Of course, participants in the policymaking process are not merely interested in getting onto the government's agenda, as many issues on the government's agenda are never acted upon. Framing an issue in a manner designed to gain the attention of a Congressional committee that is receptive to the participant's concerns and reflects a similar definition of the issue is just as important as gaining entrance to the agenda. This helps ensure that the outcome reflects the participant's interests. Congressional committees conduct hearings that help frame both the issue and the policy approach to solving it (Kingdon 1984). The elements of scope, intensity, and visibility (Cobb and Elder 1983) used to obtain access to the government's agenda must be combined in such a manner that the issue attracts participants in Congress with the same definition of the issue and some agreement on how to resolve it.

Kingdon discusses how policy entrepreneurs and an entire cast of participants such as the media, elites, elected officials, interest groups, and catastrophic events can create a sort of "primeval policy soup." Similarly, the March and Olsen (1972) garbage can model of policymaking explains how these factors all come together to help ensure that problems and solutions get coupled. To Kingdon, solutions can exist separate from problems implying that some solutions come before the problem. While Kingdon provides convincing evidence of this from the transportation and health sectors, clearly before any problems or solutions are even considered, an issue is defined. This definition process for both problems and solutions is the core of agenda dynamics and ultimately impacts policy implementation. How issues are framed in Congress is one of the key links between agenda setting and public policy.

Schattschneider also points out that how the scope of the issue is defined helps determine who participates. A policy loser, or participant who is excluded, has an incentive to redefine an issue into one that includes a broader scope to engage many of those entities who are not participating in the issue discussion. Schattschneider explains how policy losers seek to expand the scope of the issue in a manner to convert nonparticipants into understanding the issue from their perspective. The broad appeal of an issue's definition can become a "contagion" to enlisting more participants with similar interests. To Schattschneider, this increase in scope of an issue is seminal to determining which definition of the issue, and ultimately which policy, is implemented.

At the focal point of issue definition are Congressional hearings by committees and subcommittees. These hearings help define the issue as determined partially by the committees and the witnesses invited to testify. It is widely documented that with some issues monopolies may dominate policymaking in Congress. Other times, this monopoly can be disrupted and replaced with a dramatically different tone to the issue that yields a change in policymaking actors in Congress and different witnesses who participate in hearings. Baumgartner and Jones (1993) have shown that in an agenda setting process no single policy monopoly could maintain an infinite dominant equilibrium. Instead, the authors contend that there are incremental changes to the status quo that assist in creating stability in policymaking. According to Wildavsky (1984), this incrementalism approach to understanding policy change is exactly what occurs in the areas of budgeting. Dramatic reversals occur when the status quo is challenged, so that marginal alterations are not the norm during this period. Baumgartner and Jones (1993) call this stability of policymaking with periods of volatility punctuated equilibrium. How issues maintain or change definition over time clearly plays a key role in public policymaking.

Issues can take on different definitions. These definitions help shape the agenda for a particular issue that will be addressed by Congress. Often participants can mold or define an issue by associating it with a particular tone through the use of rhetoric and symbols. For instance, the nuclear industry has been used as an example of how powerful issue monopolies can control the agenda and later be completely destroyed by an opposing definition of the same issue. Early in the history of nuclear power, the issue was framed as a technological one which encouraged reliance on policymaking primarily by experts (Nelkin 1971; Nelkin and Fallows 1978), with the issue defined as national security and safety. Participants created a tone that was positive, supportive, and reinforced with a monopoly, which included one joint Congressional committee on nuclear energy and the Atomic Energy Commission, which institutionalized this definition of nuclear energy (Rhodes 1987). Today, nuclear power is a dead industry in the United States with no new plants or even the capacity for building and manufacturing nuclear plants available (Morone and Edward 1989). The change in tone of the issue from one of security to perceived risk and danger (Fischhoff et al 1984) took place over just a few decades (Weart 1988). The shift in tone can be seen in the policy outcome as well as in the dramatic changes made to Congressional committees and participants in the policymaking process. The once very positive tone shifted to a negative one with the same issue being defined in stark contrast to those institutionalized just years past. The result of this critical and negative tone was a destruction of the institutional venues in Congress and the executive branch. Not only did government venues get replaced, but the industrial ability to manufacture a nuclear plant is virtually gone in the United States. As Baumgartner and Jones point out, understanding the tone participants bring to an issue is critical because it activates the jurisdictional control of institutions and can serve as feedback to eliminate institutional structures. This is exactly what happened in nuclear energy where previously strong monopolies in Congress and industry became destroyed as a consequence of the tone changing for the issue. Companies in this market have had to refocus their work force. Few educational institutions even offer a nuclear energy curriculum or degree. Nuclear energy is one of the best examples in modern times of how definition of an issue can change and have dramatic impact in policy implementation and institutional structures.

Issue definition and agenda setting are related because changes in issue definition often lead to a different policy for consideration on the public agenda. Issue definition matters because it determines the nature of public and private mobilization efforts to encourage or discourage particular activity. This mobilization places issue definition at the center of the political debate. Stability of policy and its institutional structures can be maintained over long periods of time by two major devices: the existing structure of political institutions and the definition of issues processed by those institutions (Baumgartner and Jones 1993). This stability places issue definition at the foundation for understanding agenda dynamics in Congress. Issue definition can explain the stability of dominance by an issue monopoly and the conflict that results from the replacement of the old issue monopoly with a new one.

This issue definition process is the driving force behind agenda setting. The goal of this book is to provide a better understanding of the link between issue definitions and agenda setting in the literature. In the Baumgartner and Jones punctuated equilibrium model, the impact of this issue definition process underlies both incremental and dramatic changes in policymaking. Therefore, it is critical to understand the details of how issue definition takes place in agenda setting. As a theoretical framework, three models are used to better illustrate the issue definition process and how it impacts public policy.

Three Models of Issue Definition: Dominance, Bounded, and Valence Issues

To highlight issue definition in Congress over time, this book focuses on three models of policymaking that can take place. These three models that center on the issue definition process are classified as the dominance issue model, bounded issue model, and valence issue model. The political outcome from shaping issues using one of these three models produces distinctive policies that have different impacts on implementation. The issue definition model that forms the foundation of a policy discussion in Congress contributes to who participates in the deliberations and ultimately what agency will implement the policy. This book uses three environmental policies (wetlands, the Great Lakes, and wildlife) as cases to illustrate how each issue definition model takes shape in Congress. While the results of this book can relate to any public policy, these environmental policies were specifically chosen because they have the advantage of physically demonstrating the impact on the natural resource from how each issue was defined and shaped, then implemented. The basis for the three models is reflected in a typology, which includes characteristics such as symbols and rhetoric, Congressional committee jurisdictions and participants, redefinition of the issue, and policy implementation. These models are developed in detail in Chapter Two.

**Linking Congressional Issue Definition
with Environmental Policymaking**

This book links the areas of Congressional agenda setting and environmental policymaking. By integrating these areas in theory and applying indicators that measure the issue definition process in Congress over time. It takes a look at public policy that is clearly a different approach than previous literature. Many studies have focused on the relationship of individual Congressional members with their constituents (Bernstein 1989; Peltzman 1984). Borrowing from economics, a rational exchange type of theory called "principal-agent theory" has been used to describe the problems of viewing the public merely as merely the delegator of decisions to the agent (or elected official) which then delegates implementation to a bureaucracy (Waterman and Wood 1991; Wood 1988). These studies generally focus on the individual behavior of elected officials over time and compare the congruence and consistency of their policymaking decisions. Many rely on premises of public choice theory to understand how public policies are made by Congressional members (Shepsle and Weingast 1995; 1987). Instead of focusing on the individual level, other studies examine the behavior of committees or subcommittees (Davidson 1977; Hall and Evans 1990; Kreichbiel 1998). This includes committee assignments or jurisdictions (King 1998), and participation and nonparticipation by certain committee members (Hall 1997). A thorough literature of Congressional studies is not the point of this book; rather, this book focuses on how policy can be understood in terms of agenda setting through understanding how issue definition takes place in Congressional hearings.

The Congressional literature has not been well integrated into understanding agenda setting based on issue definition. Part of the reason for this weak link has been due to focus on the Congress or policymakers instead of the policy (Dodd and Oppenheimer 1999). Some studies have attempted to explain how Congress may influence policy through issue networks (Heclo 1979), iron triangles (Berry 1989; Cater 1964; Freeman 1965; McConnell 1967), or in political streams

(Kingdon 1984). These explanations focus on the mechanisms or structures either inside or outside Congress that can be used to influence public policy, however, the policy and how it develops from an issue is not the center of these investigations.

This discrepancy between policy and its development is particularly the problem in environmental policy literature. Many pieces have been written on specific Congressional public policymaking (which has treated the subject matter as a historical description of the behavior) by individual Congressmen. This approach makes it difficult to generalize across different policies. In addition, the subject has not been studied in a longitudinal or comparative manner but as a narrative that details the politics of passage of pieces of environmental legislation (Rosenbaum 1998, Vig and Kraft 1998, Bryner 1997). While many texts are rich in the details of specific legislation, the development of a pattern of Congressional policymaking is not made. Understanding environmental policy, particularly in Congress, has been a descriptive narrative generally beginning in the 1960s and focusing on legislative history. The main problem with beginning to explore an issue in the 1960s–1970s, especially environmental policy, is that this is the start of the modern environmental movement in the U.S. This time period does not reflect the definition of the issue prior to the pro-environmental protection tone of the 1960s. For instance, if the study of pesticides began in the 1960s, the entire shift in issue definition and participation from earlier periods of time would not be captured. This would lead to limited conclusions on how pesticides have been defined as an issue over time. Hopefully, this research demonstrates an approach that moves forward the understanding of public policy through investigation into Congressional issue definition.

Overview of the Book

Participants during a Congressional consideration of an issue (i.e. before legislation is passed), have a strong incentive to define the issue to reflect their concerns. When a policy affects many diverse interests, as is generally the case with environmental policies, this issue definition process can become strategic and tactical. If an environmental policy is framed during the agenda setting or policy formulation stage as predominately a recreational issue, for instance, the policy debate will be different than if the policy was considered a public health or agricultural issue. This same environmental issue can later be reframed to reflect another interest's concerns decades later. Therefore, environmental policy has been historically a policy area that is inherently shaped and defined by diverse groups from farmers to business and, more recently, environmental protectionists.

This book focuses on how that issue definition process takes place in Congress. How do environmental policies get defined as issues in the agenda setting and policy formulation stages? Congress is an institutional venue on the issue definition process that has not been looked at from an agenda setting approach in environmental policymaking. More important, is there an evolution of an environmental policy that can be identified where an issue becomes redefined by different interests over time? When change takes place, what is the impact not only to policy but the institutional structure connected to the previous tone of an issue? These are important questions for investigation because environmental policies clearly indicate which of the diverse, and often competing, interests got to set the agenda, and later, which interests redefined the issue to their advantage taking away that control of previous interests. Overall, this book furthers the understanding of the issue definition process in policymaking by highlighting the roles of Congress and interest groups. Understanding the issue definition process and how Congress, along with interest groups, influence agenda setting, leads the literature beyond current theories on policymaking.

To answer these questions, Congressional hearings and witnesses who testify at the hearings are important variables of study. Much of this research, therefore, places the Congressional hearing as the unit of analysis. Three models of Congressional policymaking that are proposed are evident in hearings. As mentioned before, these are dominance, bounded, and valence issue models. Each model illustrates the unique agenda dynamics associated with issue definition. Three policies dealing with the environment (wetlands, the Great Lakes, and wildlife), examined in detail from 1789 to 1999, were selected to illustrate the impact the issue definition process can visually have on a policy. Each issue represents a unique contribution to understanding environmental policymaking in Congress.

The first issue, wetlands policy, illustrates how powerful agricultural monopolies controlled the definition of the issue, which ultimately promoted the elimination of over 80 percent of natural resources since 1789 (Dahl 1991). Also, unlike the Great Lakes case, a balance of diverse interests has not occurred in wetlands policy. Eventually the issue monopoly was expanded and redefined to include issues of environmental protection similar to the Great Lakes case. Wetlands policy does not have the balance or bounded co-existence of interest as indicated by a bounded issue model. Instead, the current policy reflects a conflicting struggle where participants attempt to dominate the definition of a wetland. This struggle has resulted in a challenge to agencies required to manage wetlands. The original images of a wetland had significant impact on how the resource was defined and portrayed to the public. Wetlands were nuisance lands thought of as swamps, bogs, or wastelands. Today that image, like the policy, competes with a new definition of wetlands that acknowledges the important role wetlands have in the ecosystem. Wetlands policymaking displays what is labeled as a dominance issue model. In this model, the tone toward wetlands was creating more lands for agriculture. Development originally was institutionalized in the U.S. Department of Agricultural (USDA) policies and the Congress who subsidized drainage of wetlands. Today, there is a competing tone for wetlands which is in direct opposition to the agricultural and development tone. This alternative tone reflects a more positive definition of the value of wetlands. This has resulted in conflict because currently both definitions of the issue compete for dominance rather than work together in some form of tolerance for each other's jurisdiction. The conflict is not bounded and the issue is continually battered about in Congress. Wetlands are a source of struggle for implementing agencies like U.S. Environmental Protection Agency (EPA), Fish and Wildlife Service (FWS), USDA, and Army Corp of Engineers (ACOE). All of which have different agency perspectives on the issues surrounding wetlands policy.

The second issue, the Great Lakes policy, has an issue definition history that reflects a continued concern for commerce, transportation and economic development. Concern for the Great Lakes as a natural resource is a more recent development with strong symbols such as what has been called the "death" of Lake Erie or the infestation of exotic species. Issue monopolies associated with recreation, commerce, economic development and transportation compete for shaping the management of the Great Lakes. In addition, this issue has interesting international and intergovernmental aspects associated with it. It reflects what is labeled in the book as a bounded issue model (BIM) of Congressional policymaking. It is an example of how the definition of an issue can alter the management of a natural resource. The Great Lakes policymaking in Congress illustrates how public policies can reflect an exchange of interests that fluctuate over time while not generating tremendous conflict. The result is a balance or shared coexistence of diverse concerns such as the historic tension between multiple uses of the Great Lakes with environmental protection. This issue has trans-formed over time from a positive tone for commerce, industry and nav-igation to a more balanced or bounded one that reflects other interests such as environmental protection and public health. Policymakers have labeled their approach to the Great Lakes a sustainable development by addressing the basin as an ecosystem comprised of both natural and human resources. Before achieving that balance, the issue was redefined into an environmental protection and public health tone that was criti-cal of the previous definition of the Great Lakes policy. Today, the Great Lakes issues are bounded by the diverse participation of both tones. The once very critical tone reflected by environmental protectionists has been incorporated into the institutional venue and coexists with the interests for commerce, industry and navigation. This blending of var-ied interests had to occur if sustainable development was to be the policy definition. This is not to imply there are no conflicts or some types of cooptation. It also does not imply that these two tones coexist in harmony. Instead, the tones have not existed to the exclusion or the replacement of the other. While neither is dominant, the interests are bounded together in policymaking by separating out their tones into different venues that respect their jurisdictional boundaries. This coexistence of issue definitions has become somewhat interdependent on each other. While conflicts arise in the Great Lakes policymaking and jurisdictions are still separate based on tone, both definitions of the Great Lakes policy are mutually dependent. While it appears rational that without a healthy ecosystem of the Great Lakes, the commercial use and economic development of the natural resource would fail, many environmental policies are not defined as bounded issue or interdependent model.

The third issue, wildlife policymaking, is unlike the conflicting debate in wetlands or the coexistence of interests reflected in the Great Lakes policymaking. Wildlife enjoys the strong symbols of animals that evoke an environmental protection tone. It is difficult to mount opposition to saving animals that can be presented as going extinct such as the whales, African elephants, and others. By definition, wildlife policy was framed from the beginning with the very positive tone of environmental protection unlike wetlands and the Great Lakes. The positive tone was built on "saving" animals and "preserving or conserving" wildlife like the passenger pigeon at the turn of the century. Therefore, the issue dynamics are somewhat different because counter-mobilization is difficult. Wildlife policymaking illustrates what is known as the valence issue model since the cleavages in Congress are less likely to occur. Baumgartner and Jones found that valence policies occur with certain policies like drug, alcohol, or child abuse issues. The U.S. wildlife policy of protection was originally framed as the only legitimate tone of the issue. It remains so today despite highly publicized conflicts on certain individual animals presented in the media like the spotted owl, snail darter, or reintroduction of wolves into national parks. Trying to use symbols or rhetoric to create a negative policy image of a valence issue is virtually impossible because of how the tone was defined originally.

The three issues selected to demonstrate this typology are natural resources not pollution control policies like Superfund or hazardous waste. This typology is important because generally pollution policies are organically tied to legislation rather than issues discussed over time by Congress. These issues are used as models to illustrate how issues can be defined and redefined over time by different interests. Each policy yields a different path of Congressional politics, and more important, a different environmental policy outcome. Chapter Two explains how the three models differ in the use of images, rhetoric and symbols used to create the different outcomes. In addition, how institutional venues work, like committees that aid in the reinforcement of a particular definition of an issue, is explored. In Chapters Three through Five, the three models of issue definition are applied to the areas of wetlands, the Great Lakes, and wildlife to illustrate the linkage between issue definition and agenda setting.

The conclusions from this book show how policy definition in agenda setting can impact implementation. Wetlands policy is used as an example of the detrimental impact of a dominance issue model on agencies and implementation. The outcome has resulted in requirements for the basic definition of a wetland, confusing implementing guidance by agencies, and unclear mandates for both agency bureaucrats and wetland owners. On the other hand, the conclusions also show the positive outcomes for policies defined using a bounded issue model where ecosystem approaches with cooperating participants of an issue coexist for benefit of the resource. The improved water quality of the Great Lakes demonstrates the benefit from this issue being defined as a bounded issue model type. Finally, wildlife policies started centuries ago in the United States as protection for uses of recreation, leisure, and environmental purposes. Wildlife has benefited from the valence issue model where conflict is contained and difficult to mobilize against. In sum, practitioners as well as scholars will find using issue definition models as an approach original and informative for understanding current and past policymaking and implementation.

CHAPTER 2

Three Models of Issue Definition in Congress

Issue definition in Congress can be categorized into three distinct models: the dominance issue model, bounded issue model, and valence issue model. These issue definition models are the theoretical approach used to explain how policies can change on the agenda over time. The impact of each model on policymaking is profound when the cases of natural resources are used for testing them. Before applying the issue definition models to the specific environmental cases of wetlands, the Great Lakes, and wildlife, the theoretical characteristics of each model are described in detail. In this chapter, these models are described based on the following: the use of symbols and rhetoric, delineation of committee jurisdictions and participants, redefinition of the issue, and policy implementation. In addition, this chapter describes the research design and methods that operationalize the conceptualization of the three models of issue definition. Predictions for detecting the models in Congressional policymaking are described. The chapter concludes with some patterns that should serve as the foundation for understanding agenda dynamics. This design is then implemented in Chapters Three through Five on wetlands, the Great Lakes, and wildlife.

Linking Symbols and Rhetoric, Redefinition of Issues and Policy Implementation with Issue Definition Models

The link between issue definition in Congress and factors such as symbols and rhetoric, redefinition of issue, and policy implementation is one that is important in understanding agenda setting. Each factor plays a significant role in the outcome of a policy. These factors provide the conceptual foundation of the issue definition models.

Symbols and rhetoric are frequently used to help shape an issue for the government action (Edelman 1964; Riker 1986). The symbols and rhetoric used to define the issue involved are factors that help determine which committees have jurisdiction, which agencies are suited for implementation, and an overall direction of who has legitimacy to participate in the policy discussion. Stone (1988) describes how important symbols can be in understanding policymaking. The policy image that comes from rhetoric and symbols implies the type of discussion that Congressional committees and agencies will address. This will activate or convey legitimacy for certain participants to be involved. For instance, often policies can be formed in technological terms (to involve participants from scientific expertise) taking the public out of the realm of interest (Nelkin 1971). The result can be a policy made by a corps of technical experts who are seen as the legitimate policymakers with limited input from outside participants. Kirp (1982) argues that how issues are defined help determine the best participants for addressing the policy clearly. Symbols and rhetoric help determine the scope of who participates in policymaking.

Of course, scope of the issue is important in policymaking. For instance, Schattschneider claims that a policy loser should try to broaden the scope of the definition of an issue when it is defined very narrowly. The outcome is more participants in the issue discussion. Policy images can be used to polarize or exclude certain factions from being seen as illegitimate participants in the process. Therefore, symbols and rhetoric create the policy image that defines the issue and summons who is legitimate to participate in policymaking. Several authors have explained how policy definitions such as ones that are considered distributive or redistributive are associated with different mobilization of groups and political conflict (Ripley and Franklin 1987; 1964; Lowi 1972; Salisbury 1984).

Committees provide venues for participants to advocate their definitions of an issue. This is not to imply Congress is a passive participant. Just the opposite is true. Committees certainly have influence over the issues and clearly help define them through several means, one of the more powerful through hearings. Legislative hearings clearly create policy and require the approval of the committee for approval. These hearings can be extremely influential for understanding how the committee defines an issue. In addition, participants who testify before a committee create a record of how the issue is defined. These participants generally are selected because they represent large groups of mobilized interests. Committee jurisdictions are clearly indicated by legislative hearings. These meanings help reinforce and maintain the boundaries of a committee's jurisdiction. This is not to indicate nonlegislative hearings are not influential. Their influence is not as straightforward as legislative hearings. Nonlegislative hearings are used by committees seeking to break into new policy areas thereby beginning one tool of institutional redefinition of an issue. Committees have no restrictions on nonlegislative hearing topics. Thus, the committee often can expand its jurisdiction by holding nonlegislative hearings that add to redefining an issue (Talbert and Baumgartner 1995). When redefinition begins, the positive tone, which promotes a particular policy, is challenged over time by a more negative or critical tone. The symbols often used evoke a sense of concern, fear or loss if the status quo is maintained. This use of intensive language creates visibility of the issue to nonparticipants. Congressional committees seeking to expand their jurisdiction can adopt this redefinition of an issue along with the interest seeking to change policy. This is particularly important for participants who may also be in pursuit of redefinition of an issue and begin to "venue-shop" for a committee sympathetic to their tone of the issue, which is not, reflected in the current institutional structure. Baumgartner and Jones have coined this venue-shopping strategy as way to seek to redefine an issue by changing the tone of the symbols and rhetoric as the means toward redirecting a policy. Venue-shopping can occur in Congress by a policy loser such as an interest group or other nongovernmental participants seeking to redefine an issue based on more favorable terms.

Policy implementation is the final outcome of those policies successful enough to survive the Congressional issue definition process and the formal legislation process. Many have built typologies and models for policy implementation from the complexity of interaction (Bardach 1977; Pressman and Wildavsky 1973) to strategic formulas of success (Mazmanian and Sabatier 1983). Implementation can be mired with problems initiated from issue definition. For instance, brownfield redevelopment, which is associated with economic development, is also defined as an environmental contamination issue. As a result, the implementation of the program was assigned to U.S. EPA rather than the U.S. Department of Commerce, Housing, and Urban Development or several other agencies that could accept an economic development definition. As such, cleanup of environmental contaminants and land reuse has been the focus of implementation rather than the barriers associated with financial and economic revitalization of these plots of brownfield lands (Munro and Tzoumis 2000; Tzoumis, McMahon, and Munro 1999).

When Congressional policymaking is stable, incremental changes to policy occur without conflict over issue definition. Implementation is not negatively impacted by Congressional policymaking during incremental policymaking. However, when an issue is in the midst of a redefinition, often implementation becomes chaotic and confused. For instance, problems can result when issues are shared among agencies or when an issue completely changes tone. An issue being redefined into a tone that is not the mission of the assigned agency can significantly impact policy implementation. Implementation can become disjointed with unclear guidance and contradictory understanding of how to achieve the requirements of Congressional mandates. Thus, the factors of symbols, rhetoric, redefinition of issue, and policy implementation is what characterizes issue definition in Congress. These factors reflect an overall tone or bias which is then implemented by a government agency. Therefore, it is these three factors that distinguish the issue definition models. The next section describes.

Dominance Issue Model

The dominance issue model reflects similar characteristics of what has been termed in the literature a classical iron-triangle (Berry 1989; Cater; 1964; Freeman 1965; McConnell 1967) where a policy is controlled by a small group of participants. This cozy triangle is comprised of three participants to the exclusion of others, namely, the agency in charge of implementation, the committee in Congress primarily responsible for the policy, and the interest group that supports the policy direction. In the dominance issue model, the focus is less on these three categories of participants than how the issue dominance is maintained using symbols and rhetoric to achieve definition and policy implementation.

In the dominance issue model, symbols and rhetoric are very focused and narrow. The scope of interests is framed by how the issue is defined. Thus, the message of the symbols is one of exclusion. This means the symbols and rhetoric convey who is involved in policy and who is not. Often, the rhetoric appears to make the policy a zero-sum conflict where only certain participants' interests dominate without room on the policy agenda for other contending definitions of the issue. For instance, with nuclear energy, it is clear that opponents to it have found little common interest with proponents to it, such as the nuclear utilities. Likewise, Bosso (1987) has described the pesticides policy and, more recently, Baumgartner and Jones smoking policy, as issue monopolies that were destroyed and replaced with the exact opposing definitions of the policies. In the case of pesticides policy, it was originally defined with symbols and rhetoric that reflected the positive images of technology, progress, and more affordable and reliable food sources. The smoking policy had similar positive images of sophistication, romance, and even modernism. These images did not address concerns of public health, the environment or more negative policy images. The tone of these symbols and rhetoric helped create a policy image that offered an opposing definition. The tone of this new image is critical or negative toward the status quo. This new tone of environment and health poses a challenge to the institutional venues, which had previously excluded them.

Symbols and rhetoric can serve as strong means in defining an issue, which evokes certain images of Congressional committees and participants as having the legitimate jurisdictions for the issue. In this model, committees stake out a claim to the policy terrain, which delineates jurisdictions. These jurisdictions become stable through the institutionalization by Congress. Often, only one committee has control over the issue and literally creates the definition of an issue that reflects the committee's interests. Committees can do this through conducting legislative hearings (Talbert and Baumgartner 1995) and selecting participants in hearings that testify with the same definition of an issue as the committee's. This is an extremely stable and low-conflict situation for the participants. The committee and the participants in the hearings have a congruent tone, which becomes the dominant definition of the issue even when other definitions are clearly available. Policy can proceed in this model for decades without challenge from outside mobilized interests. Congress assigns the implementation of the policy to a federal agency that has the same definitional tone, which further reinforces the institutionalization of the policy. For instance, pesticide policy was originally assigned to be implemented by the USDA rather than any agency concerned about public health, land management with a concern for groundwater, or the environment. It is clear from USDA's mission that the agency advocates for farming interests which have the same issue definition for pesticides issues as agricultural industries. The result for decades was policy that advocated pesticides use.

The dominance issue model has an interesting redefinition process that takes place. Certainly, in this model an issue can continue for long periods of time due to this very strong reinforcing and mutually beneficial relationship among participant, and the institutionalization of the policy definition in Congress and by implementing agencies. Thus, once issue dominance is achieved, it is difficult to change. Nonetheless, change does happen when policy losers challenge the status quo definition of the issue. Change occurs through a complex process of redefinition of the issue that reflects a challenge to the structure in Congress and the implementation direction of the policy. Essentially participants change the tone of the policy by altering its definition, symbols, and rhetoric associated with the issue. Redefinition can take place in many ways. Kingdon describes how often policy solutions can be available for years as separate from policy problems. Similarly in redefinitions of an issue, it sometimes takes more than the availability of another resolution. The combination of politics, policy entrepreneurs, and issue networks needs to mobilize in some form of agreement to oppose the current definition. Media, catastrophic events like natural disasters, and other external factors assist in the mobilization. In the dominance issue model, the current definition of the issue is considered incompatible with the challenger's definition, which sets the stage for a redefinition of an issue and creates an instability not experienced during long periods of policymaking under the dominance issue model.

Redefinition of an issue under this model disrupts the stability by creating challenges to the status quo, which cannot be easily incorporated by committees or dominant participants. It usually results in intense conflict among interests, which can impact the jurisdiction and policy implementation by agencies. For instance, with pesticides policy, the U.S. EPA has been assigned a key role in policymaking that is directly at odds with the pesticide industry and often contradicts the tone of the USDA. More dramatically, nuclear energy is not an industry in the United States that has any potential for continuing, while other countries rely heavily on nuclear energy for their future needs. Likewise, smoking in the United States is now considered a public health issue rather than a recreational or leisure commodity.

This redefinition of the issue changes the tone that creates a substitution or replacement of interests and committees. Of the three models, dominance issues have the most dramatic redirection of policy that can cause havoc on policy implementation. A large-scale issue redefinition can determine the fundamental direction of public policy for decades and have significant consequences to the institutional structures that implement the policy (Baumgartner and Jones 1993). Agencies often receive confused messages for implementation by competing committees when the conflict is occurring. As this conflict unfolds, the replacement of the status quo definition with the new tone can cause a complete reversal of implementation. As described earlier, this can even lead to creation of new agencies or reassignment of the issue to agencies in the jurisdiction of different committees as was the case for pesticides, nuclear energy, and smoking policies.

In sum, the dominance issue model reflects the extremes of stability and conflict at different periods of the issue's life cycle. This life cycle is different from what Downs (1972) presented in his work on the issue of environmentalism. Certainly, over time issues proceed through a maturation process as Downs keenly points out. However, this does not mean that there is a termination of the actual issue. Instead, the issue can change definition and take on a new tone while remaining on the government's agenda, more like a reincarnation than Down's termination metaphor. In the cases of any natural resource policy, the physical resource may get redefined over time but this does indicate that there is a lack of interest or completion to the policymaking surrounding the issue. Not all policies reflect the path of long periods of stability with interruptions of great redefinitional change in policy tone. Some policies actually take a path that is less harmful to a natural resource or confusing in implementation. The bounded issue model chapter actually demonstrates how different tones of an issue and potential adversarial participants learn to coexist in policymaking.

Bounded Issue Model

Bounded issues reflect very different characteristics than that of the dominance issue model. Unlike what occurs in the dominance issue model, the symbols and rhetoric associated with bounded issues encompass rather than exclude many interests. Instead of the issue being framed and defined as a zero-sum where one definition clearly controls the agenda at the exclusion of the other definition, this model creates a boundary of tolerance where several interests coexist and influence the implementation at the same time. A bounded issue has more than one definition simultaneously institutionalized in Congress without the conflict associated with the dominance issue model. In a dominance issue model, this results in strong conflict that leads to the elimination of one of the tones of the issue. This replacement of tone and degree of conflict do not occur in the bounded issue because both definitions are interdependent with the other.

Symbols and rhetoric in the bounded issue model have less intensity at polarizing opposing interests. This is not to imply there is a consensus on the tone of the issue or even who is considered legitimate participants. The policy participants share the issue with other contenders. This tolerance or shared jurisdiction over an issue usually parses out aspects of the issue to different committees. The advantage to this approach is the shared and continuing influence over policymaking which does not lead to the risk of replacement by a contending interest. Of course, the disadvantage is often that concessions are made to a shared coexistence. Nonetheless, in this compromise model, there is an accepted tolerance of committee jurisdictions and policy images by participants advocating different issue definitions.

For a bounded issue to thrive, committee jurisdictions tend to be parsed out based on the diverse subelements of an issue. These diverse participants each have a home institutional venue, that is, a committee jurisdiction that contains their interests. The committees involved in a bounded issue have no incentive to try to polarize the symbols and rhetoric to create an issue monopoly because the conflict risks losing influence. Most hearings conducted are the legislative type. In this model, committees need not rely on nonlegislative hearings for redefinition of the issue.

Redefinition of a bounded issue does not occur because generally the issue has been parsed out in a manner within Congress and agencies to create the ability for new definitions to be incorporated into the current institutional structure. Rarely is there a large-scale conflict and most policymaking is made within this loosely diverse yet functional structure of coexistence. At first glance, one would think that this model of policymaking would have the highest potential for conflict since competition could erupt with the expansion of the issue over time. However, the stability is maintained through the decentralization of subelements of an issue. Policy can be separated out to different committee jurisdictions without substantial conflict. Different committees can simultaneously work on the different elements of the policy without encroaching on the jurisdiction of others in the policy area. The policy is defined broadly enough without problems associated with the dominance issue model during redefinition of the issue. Thus, a bounded issue rarely has large policy reversals or is ever redirected in implementation as associated with the dominance issue model. Policymakers and implementers benefit under this model. While there is decentralization to different subcommittees in Congress, there is not a reversal in policy definition. The agency staff may feel some strain in pulling together an implementation strategy that has to coordinate with multiple agencies and subcommittees but this is far superior to the problems associated with implementation under the dominance issue model.

Unlike the dominance or bounded issue models, a third type exists, which is somewhat rare, called the valence issue model. This unique model of policymaking is monotonic by definition. The next section describes this model.

Valence Issue Model

Valence issues are the most unique model of policymaking particularly in the area of symbols and rhetoric. These issues take on images through rhetoric and symbols that are different from the previous models. In a valence issue model, countermobilization to the status quo does not occur due to the powerful symbols that initially defined the issue. The symbols and rhetoric are not powerful because they have a tolerance that can encompass many interests as with the bounded issue. Just the opposite is the case. Instead, the valence issue is overwhelmingly effective at neutralizing any other definitions of the issue because of the credible and legitimate messages it evokes. For instance, no one would claim to be against the protection of children against abuse, or a proponent of drugs, illiteracy, homelessness, or other issues that evoke no opposing definition. Baumgartner and Jones found drug, alcohol, and child abuse to be good examples of how these valence issues can occur in policymaking. Throughout the years, these issues have had different levels of salience; the issues remain generally popular with the public.

Valence issues are dependent on how the definition of the issue was originally introduced on the agenda, resulting in committee jurisdictions that are stable. Typically, only a few committees dominate the policy. Policymaking occurs in Congress primarily through legislative hearings. Because challenges to jurisdictions are rare, the use of nonlegislative hearings should be less than the legislative type. Like the bounded issue model, implementation agencies do not have to deal with large redefinition of the issue or policy reversals of the dominance issue model. They also do not have to be concerned with committees giving them conflicting direction or coordinating among the diverse interests involved with a bounded issue.

The valence model explains how policies can continue on a course of manageable conflict while maintaining some consensus for support by the public. In the valence model, issues generally can withstand any challenges made to the definition. Generally, these challenges tend to be more minor and often based on some localized conflict which does not impact the overall symbols, rhetoric, and ultimately, the tone of the issue. No policymaker would disagree with trying to stop child abuse in the United States. However, how policies against child abuse are implemented is another matter. There may be many different solutions, but at no time does consensus ever waiver on the tone of the policy.

Table 2.1 summarizes how the three models of issue definition are characterized by the factors of symbols and rhetoric, redefinition of the issue, and policy implementation. These three models reflect how issues can take different paths to policymaking. Often issues stay contained in a single policy type, which is reinforced by the institutional structure of Congress and implementing agencies, as well as the definition of the issue. It is not usually the case that an issue can simultaneously reflect more than one type of policymaking based on this typology.

Table 2.1 **Conceptualization of Issue Definition Models:**
Issue Definition Models

Characteristics	Dominance	Bounded	Valence
Symbols/Rhetoric	Very focused and narrow scope of a policy image which excludes others and is polarizing. The policy image created serves to mobilize and countermobilize interests.	Encompasses many interests. Creates multiple policy images of the issue.	Very powerful neutralizing policy images that evoke legitimacy of the status quo.
Redefinition of Issue	Frequently reflects stability with lack of conflict during periods of dominance. Redefinition of issue usually indicated during high conflict with substitution of a new dominant policy often comprised of those previously excluded.	Often reflects contained conflict with large scale redefinitions of issue not likely. When redefinition occurs, the outcome is tolerance of simultaneous definitions of issue co-existing over time.	Manageable conflict. Significant redefinition of issue rarely takes place.
Policy Implementation	Can have a dramatic impact and redirection on policy implementation.	Moderate impact on implementation. Redirections in implementation rarely occur.	No significant impact to policy made. Changes are incremental and create stable implementation.

Thus, this approach easily lends itself to testing by sampling issues for the patterned characteristics described. Table 2.1 serves as the conceptualization for how the models will be tested. The next section describes in detail how the models will be operationalized for testing.

Testing the Policy Models of Issue Definition

This section outlines the fundamental design and methods that are used to test the policy models of issue definitions and the relationships described between Congress and participants, and the tone of the issue. To test these models, clearly the role of tone is a concept that needs to be operationalized. Tone reflects the definition of an issue. To identify change, a baseline tone must be established. When new definitions of the tone challenge the status quo, a separate tone is established which is captured by using a developed coding scheme. The following sections describe how this is accomplished through a dichotomous system for coding tone. In addition, the following section outlines how tracking participants, committees, and hearings over time can be used to identify the three issue models. The next section describes predictions made for identifying three types of issue definition models. The actual tests of these predictions are made in Chapters three to five.

**Coding Tone of Hearings, Committees,
and Participants Who Testify**

One of the main factors in detecting how issues are defined is looking at the tone of the issue. One of the best methods to understanding how the issue is framed is to identify the tone of issues being considered on the Congressional agenda. Tone was coded for committees, hearings, and participants who testified (witnesses). Using the electronic versions of Congressional hearings from 1789 to 1999, the hearings were extracted from the Congressional Information Service (CIS) on CD-ROM. CIS assigns key words that describe the content of the hearing. This is done in a multiple assignment manner. That is, one hearing may have many key words indexed by CIS. This is important because it ensures that hearings are not excluded when searching for a key word. Hearings were searched using the key words the Great Lakes, wetlands, and wildlife with the start of the Congress. but a search would capture any use of the word in the indexed key words provided by CIS. In addition, CIS provides synonyms of these words as part of their key words associated with each hearing. A hearing would be selected if it contained swamps, bogs, and marshes in addition to wetlands. Likewise, any of the five individual Great Lakes names would be captured in addition to the proper title the Great Lakes. For wildlife, synonyms included endangered species, threatened or extinct species, conservation and wildlife management. The search included not only these key words that CIS assigned to every hearing, but also was programmed to capture any witness name, title of organization, hearing content, committee or subcommittee name, or description of witness testimony that included the keywords or synonyms. This search process was the broadest possible to ensure a complete sample was obtained from the population of hearings.

These Congressional hearings from 1789 to 1999 were used to look at the longitudinal aspect of how issues can be defined and redefined in Congressional hearings over time. The time frame encompasses the entire life cycle of an issue in Congress. This approach yields more valid information than a cross-sectional approach focused on only a few years. This is particularly important for environmental issues that often evaluate an issue beginning in the 1960s. Generally, this would yield a very skewed result of an issue, which would only reflect a more recent definition and tone of the issue.

The sample cases of wetlands, the Great Lakes, and wildlife policies were selected for testing these models for several reasons. Because so much of the environmental policy literature has limited scope in understanding Congressional policymaking, this contributes significantly to both agenda setting and environmental policymaking. Also, within the realm of environmental policies, natural resources were selected versus pollution policies. The importance of choosing several natural resources as issues to explore prevents biases that come from tracking technology discoveries not yet made in the 1800s. If nuclear policy were examined as an issue, the cycle of issue definition would be limited to more recent decades since nuclear reactions were not discovered until the 1940s. If cleanup of waste was to be the selected issue, it would be limited to tracking Superfund legislation rather than the issue of waste. The same problem exists for recycling hazardous waste. Thus, this research has sampled the three issues of the Great Lakes, wetlands, and wildlife. All three of these natural resources existed prior to legislation even though they may have been labeled in different terms, e.g., the word wetlands may not have been used until the 1960s, but the term bog, swamp and other synonyms were certainly used.

Tone was operationalized into a dichotomous coding scheme that has been tested in the literature (Baumgartner and Jones 1993; Jones et al 1993; Tzoumis 1999; Weart; 1988). This dichotomous coding scheme was used for all three environmental issues with some adaptation based on the issue. There should be at least one definition of an issue over time. The original tone was considered the status quo definition which reflected the original established definition of the issue used as a comparison. Criticism or challenges to that original tone were coded as negative to the status quo. These challengers had new definitions of the issue that were not incorporated into the status quo. A third code for neutral or uncodeable was used to ensure that all hearings and participants were included. This was often the case for hearings involving appropriations that can serve as formalities in funding.

This binary system of status quo versus challenger definitions for coding tone is a valid approach to analyzing the entire life span of an issue. For instance, if an issue monopoly exists the tone should monochromatic, reflecting a dominance pattern. Because this research looks at the entire life span of the issue in Congress, a dominance issue should be challenged over time to reflect high conflict in the hearings and implementation. Abrupt changes in tone should result from the outcome of conflict in the pattern over time of a dominance issue model. Thus, baseline tone, that is, one supportive of the original definition of an issue is recorded. Any deviation of that tone of issue definition is recorded as a negative since it is considered an alternative or challenging definition to the definition. In the dominance issue model, the pattern over time should be some significant duration of dominance by the baseline tone. Redefinition of the issue is indicated by the challenging critical or negative tone that replaces the original tone. The words positive or negative merely refer to a baseline versus change. This designation of the variable tone is not based on opinion or judgment of the issue.

When there is a bounded issue model, the pattern is very different. Both tones are reflected in Congressional policymaking for a long period of time. The tone of the issue does not necessarily change, but the two tones coexist without conflict and negative impacts to implementation. The dominance issue model has the pattern over a long period of time, and the definition of the issue dominates. A bounded issue has a tone within the same year divided between baseline and challenging definitions of the issue. The baseline definition remains over time and shares agenda setting with another tone. The change in tone of a bounded issue does not eliminate the original baseline definition as with the dominance issue model.

The valence issue model pattern for tone would be similar to the dominance issue model during the periods of stability. In the valence issue, we would not see challenges that come from redefinition as in the dominance issue model. The issue would not be plagued with the chaotic implementation impact that comes from the high conflict with a dominance issue model. The valence model reflects a consistent pattern over time for one tone without any redefinition. Committee jurisdictions are established and participants reoccurring in hearings.

The baseline tone was established for each of the individual issues examined in this research. In the case of wetlands, original tone of the issue was defined as representing agricultural and development interests. Based on the literature on wetlands, this was the original tone of the issue in Congress so it was used as a baseline. Thus, this original tone reflects issues that are favorable toward farming, private property owners, and developers. Early in the history of many countries, wetlands were thought of as nuisance lands. The main definition of the issue was framed by having to develop these lands. The opposite of this tone is environmental protection. Therefore, the second value for coding this tone was defined as environmental interests, which can include preservationism, water quality and flood management, sport, leisure and recreation. This challenging tone framed the wetlands as valuable. It is not important why the wetlands are argued as valuable since the outcome in implementation and policy would be the same, that is one of environmental protection of the natural resources. Again, the history of wetlands science and policy support the binary system of coding employed here. The original or baseline tone reflects the destruction of wetlands; the challenging tone reflects the protection of wetlands. The reason for protecting or destroying wetlands does not matter; the definition of the natural resource as being protected or eliminated does.

In the case of the Great Lakes, the original tone based on the historical literature is focused on industry, commerce, and navigation. How to use the Lakes for industry, commerce and navigation for expansion of a growing country was inherently the original baseline issue definition of the Great Lakes. A challenging tone to this would be the protection of those Great Lakes. Thus, environmental protection of the Lakes was the opposing tone that reflected a redefinition challenge to the baseline. Like wetlands, this environmental protection tone included interests of recreation such as leisure activities of boating, fishing, and water sports along with environmental preservation of the resource. Again, like wetlands, the reasons for the tone are not important. In fact, often tone was defined by a collection of interests that formed coalitions. Therefore, the two tone approach to understanding the Great Lakes policies is the best methodological design toward understanding the patterns of the three models that are being investigated.

The original tone was the protection of the species by interests such as animal rights activists, environmental preservationists, recreationalists and leisure users of wildlife. This tone includes the interests of preserving wildlife for human use but the focus is on limiting consumption. In the case of wildlife, one definition is divided into commerce, trade and industries associated with the selling of rare or exotic species or the displacement of wildlife. It reflects the same dichotomous tone for coding as with the Great Lakes and wetlands. This challenging tone for wildlife involves the sale of native species (nonhusbandry species) in a commercial manner and the use of wildlife in industry and development. The original protection policies for wildlife are reflected in the early historical literature and should be reflected in discussions in Congress as the original tone. Included in this baseline tone as well are the use of native wildlife in the United States and other countries for ornamental products, manufacturing goods, and any consumption of the wildlife.

In a valence issue model, it is expected that tone would mainly endure the life span of the policy. To test for this endurance, coding of wildlife for both tones was made. It was expected that a majority of witnesses, committees, and hearings would reflect the protectionist tone since it was considered in the historical literature as the original tone. Unlike the Great Lakes and wetlands, wildlife was first defined as needing protection from commercial uses because of the history of wildlife and U.S. growth. Wildlife was recognized as needing government protection. Unlike the other two issue models, the valence model may have a challenging tone to baseline, but the baseline tone survives any attempt at redefinition over time. Moreover, a challenging tone is difficult to mount against a valence issue.

Overall, tone was one of the key variables coded for all hearings, committees, and witnesses. In sum, a dichotomous variable was used to establish an original baseline tone to compare with a challenging one. CIS provided summary information on the purpose of the hearing, the committee holding the hearing, and the testimony of witnesses. This method was used to assign tone codes over time. A similar method for coding tone can be found in Baumgartner, Jones, and McLeod (1998).

Tracking Who Participates in Congress

Every committee, subcommittee, and witness who testified was given a specific identifier code to identify its pattern of participation. Different patterns were associated with each of the issue definition models. Reoccurrence of witnesses or committees, for example, was expected in the dominance and valence issue models at different points in time. Both models were expected to be indicated by frequent participation of interests with the dominant tone of the issue. This would be triangulated in the hearings, committees, subcommittees, and witnesses who testify. If dominance was occurring, the tone of the hearing, committee, subcommittee, and witness should be congruent. In addition to coding the tone of the witness, many other variables were included. Witnesses were coded for type of organization they represented such as industry, trade association, interest group, elected official, international country, federal or state/local agencies, or individual citizen. Gender, location of witness, and chamber of testimony were also included. Patterns based on the character of the three issue models were then identified based on predictions shown in Table 2.2.

Table 2.2 Predictions for Models of Issue Definition:
Issue Definition Models

Indicators	Dominance	Bounded	Valence
Tone	One tone dominates until another replaces it.	A baseline tone and challenging tone coexist.	One tone occurs with limited challenges.
Hearing Type (Redefinition tool)	Nonlegislative hearings are used by challenging tone during periods of re-definition.	Legislative hearings.	Legislative hearings.
Number of Hearings (Attention)	Few hearings until periods of re-definition.	Many hearings.	Few Hearings.
Number of Committees (Venues)	Very few committees until periods of re-definition.	Many committees and subcommittes.	Few committees.
Participants (Witnesses)	Re-occurring participants from narrowly defined sectors with similar tone as dominant interests until periods of re-definition.	Diverse participants from different organizations and sectors.	Re-occurring participants with heavy participation from government sectors.

A bounded issue model is indicated by a diverse mix of tone for hearings with no single tone dominating. No single tone for committees, subcommittees, or witness would dominate. Likewise, there is no exclusionary pattern of jurisdictional control so several committees with different tones conducted hearings. This represents a stable pattern where a larger mix of witnesses from a variety of represented interests participated. For a dominance issue model, this pattern would not last; for the valence issue model it would be the only pattern over time. Participation in a valence issue model includes heavy reliance on government participants like environmental agencies and environmental committees. Dominance issue models display new participants replacing previously reoccurring ones after issue redefinition is completed.

In addition to tone, hearings are coded as legislative or nonlegislative. Hearings that consider bill-referrals are defined as legislative. Generally, committees tend to use nonlegislative hearings to encroach on the jurisdictional control of another committee that would indicate a challenge to a committee's monopoly. Nonlegislative hearings are particularly important because few restrictions exist on the subject of hearings. Talbert and Baumgartner (1995) showed how Congressional committees use them to redefine policies in order to expand their jurisdiction and to force rival committees to act on matters that they might avoid. Thus, these nonlegislative hearings should occur more frequently during periods of issue redefinition.

It was anticipated that more legislative hearings would be indicated in a dominance model with a period of nonlegislative hearings to redefine the issue. However, the pattern would resume to legislative hearings once the new issue definition monopoly was established in place. Just the opposite would be indicated for a bounded issue model. There, it was expected that both types of hearings would continue for the duration of the issue. The valence issue would have only legislative hearings without any challenge of the nonlegislative hearings. In this model, the nonlegislative hearing was not a useful policymaking tool because the issue is, at best, incrementally redefined if at all.

By tracking tone, hearing type, and number of hearings, committees and participants, the patterns of which issue definition process directs the agenda is exposed. Thus, tone provides the link between issue definition and agenda setting. This data, summarized in Table 2.2, was used to test the relationships expected to occur within each model.

Testing the Relationships within Issue Definition Models

Congressional policy dynamics have a significant role in helping promote definition of an issue. Congressional committees can do this through three main routes or venues. First, committees influence the definition or redefinition of a policy by conducting hearings with a certain tone. That is, instead of considering issues of the opposing tone, committees only conduct hearings that reflect their tone. Usually, the topic being considered is not neutral and the subject of the hearing is not open to consider a wide range of alternatives. Second, witnesses who testify before a committee can either reflect the tone of the committee or an opposing tone. Committees concerned about maintaining or expanding their jurisdiction have an incentive to select witnesses that have a corresponding tone. These two dynamics actually reinforce each other by satisfying the needs of policy advocates who need to venue-shop because their interests are not being considered by the status quo in Congress and by committees seeking to expand their jurisdiction. To investigate if these Congressional dynamics exist in wetlands, the Great Lakes, and wildlife issues, a correlation should occur between tone of the hearings and committees, and between tone of the committees and the witnesses. To investigate venue-shopping and Congressional expansion, two tests were conducted using Chi-Square and correlation coefficients of Gamma and Kendall's Tau-b. These statistical tools will indicate if there is a significant relationship and the strength of that relationship. Because the variable is binary (though when the neutral category is included it could be considered ordinal) the analyses also included both nominal and ordinal measures of that relationship by using both Gamma and Kendall's Tau-b.

The third Congressional dynamic that impacts the three issue definition models is the relationship between Congressional attention and venues. As Congressional attention on these three issues increases, the number of venues claiming jurisdiction also should increase. Congressional attention to wetlands, the Great Lakes, and wildlife attracts new venues to capture jurisdictions that are unclaimed by others. An ordinary least squares regression, used to test how much Congressional attention is paid to these three cases, explains the growth of venues claiming the jurisdiction. Equation 2.1 shows the ordinary least squares equation used in this analysis.

The predicted relationship: Congressional attention measured as the number of hearings on the particular environmental issue per year (independent variable) results in increases in the number of available venues as measured by the number of committees (dependent variable). An endogenous lag variable was included to control for hearings conducted the previous year on the particular environmental issue. This lag variable allows the regression model to reflect actual growth in Congressional attention that cannot be explained by the previous year's hearing activity.

Equation 2.1 Congressional Attention and the Role of Committees

$$Y_t = \beta_1 X_t + \beta_2 Y_{t-1} + e_t$$

Y_t = Number of Hearings conducted each year on the environmental issue (dependent variable).

X_t = Congressional Committees holding Hearings on the environmental issue each year (independent variable).

Y_{t-1} = Number of Hearings conducted previous year on the environmental issue (lag variable).

e_t = residual unexplained by model.

Chapters Three through Five provide detail analysis of the three issue definition models. These chapters apply the concepts outlined here and provide the results for each issue. Comparisons across the three models are made in the final chapter.

CHAPTER 3

Dominance Issue Model—
The Case of Wetlands

There is a long history of policymaking in Congress that has a significant impact on the management of wetlands as a natural resource. As such, the importance of understanding how Congressional committees define wetlands policy influences, if not totally determines, how wetlands are managed. This chapter looks at the issues and participants involved in Congressional hearings dealing with wetlands policy from 1789 to 1999. Results show that three major policy eras in Congressional hearings occurred over time that have significantly impacted the management of wetlands as a natural resource. The research describes how these changes in policymaking took place and how issues of agriculture, science, and public awareness about the environment aided in the policy changes. This chapter describes how wetlands policymaking in Congress has changed over time. Analysis using statistical tests that include tone of witnesses, hearings and venues lend evidence to how increased Congressional attention can lead to increased venues for wetlands policymaking in Congress. The findings explain how wetlands resemble a dominance issue model, which continues to experience significant conflict.

Relationship in Wetlands Policymaking
of Congressional Hearings and Committee Dominance

Congressional wetlands policymaking in the United States has evolved over time. From examining Congressional hearings on wetlands, three distinct eras can be seen. The early years of wetlands policymaking in

Congress range from 1789 to 1945. This first era was dominated by the needs for agriculture and developing land for a growing country. The second era from 1946 to 1965 reflects a period of declining influence of the agriculture and development issues over wetlands policymaking. Finally, the third era which begins in 1966 and continues to today, is characterized by the tremendous conflict between the participants and issues that dominated in Era I and those that emerged in Era II.

Using the Congressional Information Service (CIS) on CD-ROM for Congressional hearings from 1789 to 1999, a search was performed using the keyword wetlands. To avoid missing hearings on wetlands before the term was commonly used, synonyms such as bogs, marshes and swamps were included by CIS under the term wetlands. This search yielded a total of 248 hearings with 1580 witnesses and reflects the complete set of Congressional hearings on wetlands policy in the United States.

The tone of committees, witnesses, and hearings were recorded as a dichotomous variable as described in Chapter Two. For wetlands policy, tone was defined as either representing agricultural and development issues or environmental protection issues. For coding environmental protection tone, issues included concerns about protecting and preserving wetlands as a natural resource. Often the environmental protection tone included environmental interest groups and sometimes recreational groups interested in hunting, sports and leisure. For coding agricultural and development tone, issues included being favorable toward farmers, private property owners, and developers. Since the divisions among these groups in the wetlands policy debate are distinct, it was a valid classification to make in the coding effort. Hearings and witnesses were coded separate from the committee based on the information provided by CIS. This technique allowed comparisons between tone of the venue and tone of the hearing and witnesses. Key issues of concern by witnesses were also tracked over time.

Congress made wetlands policy using twenty-eight committees and forty-seven subcommittees. Figure 3.1 shows the total of annual hearings on wetlands policy. No hearings were conducted before 1847. One committee held hearings in 1847 and 1849. Figure 3.2 shows the

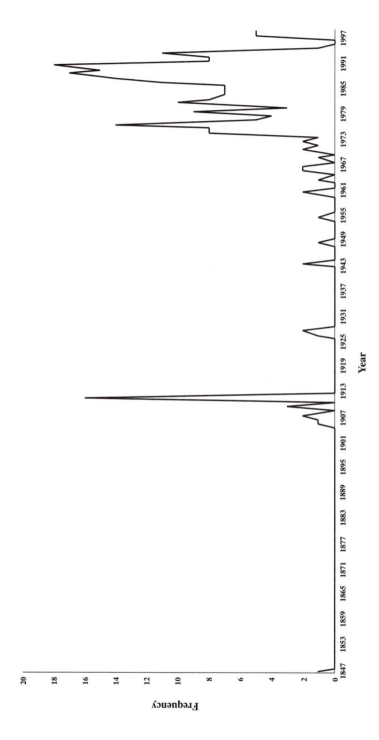

Figure 3.1 Congressional Hearings on Wetlands

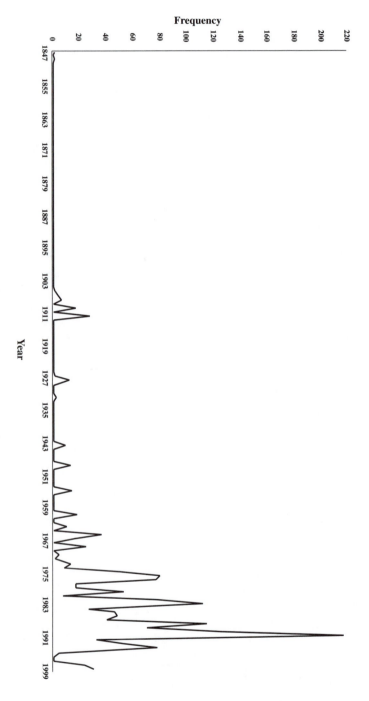

Figure 3.2 Witnesses who Testify on Wetlands

number of witnesses that testified annually on wetlands policy. Both of these data sets are used for the analysis of wetlands policy (less than five witnesses testified in 1847 and 1849 which are not on the chart). In general, Figures 3.1 and 3.2 show a tremendous increase in overall hearings and witnesses who testify before Congress beginning in Era III. Besides tone, this information was coded for supplemental data information, such as gender of witness and organizations represented. Individual codes for each witness were used to track if witnesses testified multiple times indicating a reliance on certain interests by Congress.

In Figure 3.1, it appears there was very little interest by Congress on wetlands until the late 1960s. Witnesses who testify are also limited until the early 1970s where Figure 3.2 illustrates a large number of witnesses appearing before Congress. The next section on results describes in detail the three eras in Congressional hearings on wetlands.

Results

Three Eras of Congressional Policymaking on Wetlands

By examining the annual percentage of hearings in the environmental protection venue versus the agricultural and development venue, several trends emerge. Over time, three eras can be identified in wetlands policy based on the percentage of hearings using one of the two tones over time. There are the agricultural and development dominance years from 1847 to 1945, the early challenges to dominance from 1946 to 1965, and the loss of dominance beginning in 1966 to the present. Table 3.1 shows the data from these three eras used to analyze the overall trends in policymaking. For each era, Table 3.1 breaks down the venue and witness information presented in the following section.

Era I—The Agricultural and Development Years (1789–1945)

Early in our history, wetlands were not thought of as a beneficial natural resource. Mitsch and Gosselink (1993:12) describe how wetlands were depicted early in United States history as useless land that was more of a public nuisance than a valued natural resource. It is estimated that in the 1780s, land that now comprises the United States originally contained 392 million acres of wetlands (Dahl 1990). This was considered as poor quality land that needed to be drained, cultivated and made useful for agriculture or development. Approximately half of the wetlands in the continental United States was lost between the late 1700s and the mid-1980s (Dahl 1990). Although there are some problems with comparing estimates of wetlands loss (especially over time) since studies use different definitions of wetlands another study estimates that by 1954, drainage, fill, and construction had destroyed almost forty percent of the nation's wetlands (Kuslar 1983). According to the Clinton Administration Wetlands Plan (1995) only about one hundred million acres of wetlands remain which represents less than five percent of the landmass of the continental United States. In all the estimates, clearly a significant loss of wetlands has occurred in the United States.

Conversion of swamps into lands considered productive for a growing country was how wetlands policy was framed in these early years. In fact, the term wetlands came into use only in the second half of the twentieth century (National Research Council 1995). Since the mid-1800s, Congress encouraged the conversion of wetlands to agricultural use primarily for farming crops. This practice of conversion was the dominant policy of managing wetlands. There was general acceptance of this approach to wetlands management. In fact, Congress conducted no hearings before 1847 on wetlands. There was no debate or discussion on the preferred management practice of draining wetlands for a more productive land use. The first Congressional hearings begin in 1847 and continue in 1849. These hearings directed the sale of marsh land in Indiana. Congress began a systematic sale of wetlands with the Swamp Land Acts of 1849, 1850, and 1860. During this time, wetlands were considered a hindrance for urban growth and agricultural progress. It was not until 1929 with the Migratory Bird Conservation Act and the 1934 Migratory Bird Hunting Stamp Act that concern for decreasing migratory bird populations and their habitats made the issues of wetlands to be considered potentially beneficial. However, witness testimony showed that preservation or protection of the wetland itself was not the primary concern. Moreover, in the 1930s the USDA continued to subsidize the drainage of wetlands for agricultural production.

Several indicators in Table 3.1 show the strong monopoly that promoted agricultural uses for wetlands in Congress from 1789 to 1945. For instance, of the twenty-eight hearings held, primarily agricultural and development committees conducted them. The four committees that conducted the hearings were included the Committee on Agriculture and Forestry, Public Lands, Indian Affairs, and Expenditures in the Department of Agriculture. Most of the hearings were nonlegislative (68%) with the majority of hearings conducted by agriculturally-related committees (68%). This was clearly an era of dominance by the agricultural and development committees over policymaking.

Witnesses represented a narrow range of interests such as commerce and business (2.9%) and environment (7%) with the largest representation from agriculture (33%). Not surprisingly, the major issue mentioned by the witnesses (39%) in this era focused on the key issue of wetland purchase and land transfers. Interestingly, federal agencies (24%) and federal elected officials (21%) dominated the representation of witnesses. These witnesses generally sought the support of Congress to develop the land by removing the perceived swampy areas. However, no single witness dominated in testifying before Congress when repeat appearances were tracked. In fact, most witnesses appeared less than two percent of the time in this era. The number of legislative hearings (32%) was approximately one-half of nonlegislative hearings (68%). This is surprising because based on the predictions made in Chapter Two, most dominance periods are characterized by legislative hearing periods. Nonetheless, during this era of dominance, agricultural and development committees tended to use nonlegislative hearings to control the agenda on wetlands. The geographical areas of witnesses that dominated the testimony were split among the Midwest (23%), the South (29%), and experts from the Washington, D.C. area (30%). Geographical tracking of witnesses is important because clearly wetlands are only located in certain regions of the United States. It is not a surprise that participants in hearings come from the Midwest and South where the majority of wetlands that impact agricultural interests occur.

Table 3.1 Congressional Wetlands Policymaking from 1789–1999

	Era I (1789–1945) Agricultural and Development	Era II (1946–1966) Environmental Protection	Era III (1967–1999) Conflict
A. Hearing Information			
Number of Hearings	29	8	212
Number of Committees	4	5	21
Number of Subcommittees	0	3	44
Percentage of hearings *held by each Committee*			
Agriculture Related	68%	14%	< 10%
Merchant Marines and Fisheries Committee	0%	29%	18%
Public Lands	21%	0%	0%
Conservation Wildlife	0%	14%	5%
Percentage of hearings held *by each Subcommittee*			
Fish and Wildlife	0		All < 12%
Conservation		42%	
Merchant Marine and Fisheries		14%	
Type of Hearing			
Legislative	32%	71%	62%
Nonlegislative	68%	29%	38%
Type of Committees			
Commerce	0%	14%	6%
Environmental	21%	57%	52%
Agricultural	68%	29%	10%
B. Witness Information			
Number of Witnesses	70	64	1446
Gender of Witnesses			
Male	52	61	1228
Female	1	0	164
Uncodeable	17	3	49
Percentage of Witnesses *that testify before each Committee*			
Merchant Marines and Fisheries	0	42%	21%
Agricultural Related	52%	All < 10%	All <8%
Environment and Public Works	NA	0	20%
Public Works and Transportation	NA	0	15%
Percentage of witnesses Testifying *before each Subcommittee*			
Fish and Wildlife Conservation	0	14%	All <15%
Merchant Marine and Fisheries	0	14%	
Percentage of Selected Witnesses			
Federal Agencies	24%	27%	25%
Federal Elected Officials	21%	30%	7%
Interest Groups	1%	1%	20%

Percentages are calculated based on each era. Only major percentages are presented. All categories for each variable are not included because they reflected a smaller percentage.

These multiple indicators clearly demonstrate the dominance by agricultural and development interests in wetlands policymaking. Both venue and witness indicators presented in Table 3.1 confirm the existence of this monopoly. The weak representation of environmental protection as a tone in either testimony by witnesses or in a hearing conducted by a committee reinforces the interpretation that a consensual and closed monopoly took control of wetlands policy in Era I. Thus, during this period there was little challenge to this definition of how wetlands were to be managed as a natural resource during this period.

Era II—Early Challenge to Agricultural Dominance (1946–1965)

A change in wetlands policy from Era I takes place during 1946 through 1965. This policy change is characterized by a gradual decline of the dominance of agricultural and development issues for wetlands policy. Dennison and Berry (1993:14) describe three socioeconomic factors that impacted this change in wetlands policymaking. After World War II, the United States experienced large gains in economic prosperity for a growing middle class. Median family incomes rose in the early part of this era which assisted in a rise of expenditures for housing and consumerables. The first of the socioeconomic factors impacting wetlands policy was a tremendous migration from urban centers to often undeveloped areas outside of cities. Besides seeking improved housing by the middle class, the growth of the population in the 1950s added to the need for more housing. As a result, many suburbs developed to meet the desire for new housing. Along with this move to the suburbs came development of land due to increased demands for infrastructure like roads, highways, retail, education, and recreation. Most of this growth took place outside of large cities where many wetlands still existed. These socioeconomic factors played a role in the policy shift seen in Congressional hearings from Era I to Era II.

Second, wetlands science was beginning to mature into a highly technical field from the more descriptive one of the previous era. Prior to the 1960s, few studies examined wetlands in detail. However, by the mid-1970s, wetlands became the focus of much investigation (Mitsch and Gosselink 1993) and continue to be so today. In Congress, there was an emergence of new environmental interests in wetlands that did not stem from the agricultural and development monopoly. Scientists began to develop technologies better able to identify the impact of wetland losses on ecosystems. The investigation into the scientific functions and human values of wetlands underwent significant growth in this era. Nonetheless, scientists were beginning to create technical definitions of wetlands which later would cause major controversy with regulators in Era III.

Third, the 1960s brought a public awareness of environmental issues. There was a tremendous growth in public concern and interest group activity about the environment (Dunlap and Mertig 1993). The media and public opinion focused the Congressional agenda on environmental problems associated with air pollution, contamination of water, and problems with pesticide use. By the 1970s, the public concern of the 1960s was codified into several pieces of major legislation that continue to frame environmental policy today.

Era II reflects a time when agricultural and development issues are not the only influence on Congressional policymaking on wetlands. However, support for utilizing wetlands for agriculture and development continued to exist in Congress. While they were no longer the only issues considered, agricultural interests still retained their powerful Congressional venues. In fact, from 1954 to 1974, there was an intensive period of development and wetland conversion with approximately ninety-five million acres of wetlands being lost (Dennison and Berry, 1993; 67). Thus, this decline of the agricultural monopoly is not a destruction or elimination of the interest, but more of a challenge to its sole influence over wetlands policy in Congressional agenda setting.

Beginning in the mid-1940s, while wetlands policy in Congress appeared to remain as noncontroversial as in earlier years with few hearings conducted, a closer examination of the committees and subcommittees showed an increase in participation by environmental protection interests and a change in tone of key issues discussed in witness testimony. The committees in this period include Conservation Wildlife,

Merchant Marine and Fisheries, Commerce, Irrigation and Reclamation, and Agriculture. In this era, the majority of the hearings are conducted by the Merchant Marine and Fisheries Committee (29%). Unlike the previous era, these committees begin to expand their jurisdictions into the monopoly held by agricultural interests. New venues are opened to environmental interests that reflect the beginning of a redefinition of the wetlands issue away from the agricultural tone that previously dominated. New venues are available for the policy losers who are not supportive of agricultural and development issues. Even with this opportunity of new venues, however, very few hearings (8) were conducted. But, of the eight hearings, only two were nonlegislative (29%), unlike the prevalence of nonlegislative hearings (71%) in the previous era. Table 3.1 shows the mixture of agricultural and development and environmental protection venues beginning to compete during this period.

Most witnesses testified before the Committee on Merchant Marines and Fisheries (42%), with only ten percent of the witnesses testifying before the Agricultural Committee. In addition, the Subcommittee on Fisheries and Wildlife Conservation heard the most witnesses (14%). The break up of the monopoly was underway. The majority of witnesses represented either federal agencies (27%) or were federal elected officials (30%). Thus, the federal government continues to participate as the largest group represented in hearings as witnesses. However, like Era I, no single witness dominated wetlands policymaking. The largest majority of witnesses (38%) came geographically from Washington, D.C. This is different from Era I when witnesses came from geographical areas with major wetlands. While witnesses from Washington, D.C usually participated in Era I, with the largest percentage, now they dominated. Clearly, no longer did agricultural interests from the Midwest and South have to testify before Congress to drain wetlands. The environmental protection interests were now mobilized in Washington, D.C. The key issues now discussed by witnesses tended to focus more on the protection of the wetlands (54%) and sport/leisure (22%). Unlike Era I, no witness discussed the transfer or sale of wetlands. The results from this era clearly show the early entrance of an environmental protection tone into the policymaking considerations for wetlands management. New committees and witnesses participate in hearings that reflect a different tone than in Era I. As this era closes, it serves as the precursor to the highly conflictual debate in Congress that begins in Era III.

Era III—The Loss of Dominance of Wetlands Policymaking (1966–1999)

The growing sophistication of wetlands science and environmental organizations in the 1970s and 1980s, along with the focus of proving economic benefits in a cost benefit analysis for environmental policies in the 1980s, added to a clash of interests. The ecological values and functions of wetlands became strongly recognized in the 1970s. In addition, there were many studies on the economic valuation of environmental resources using modeling and "willingness to pay" schemes during the 1980s. But it was the differences in technical definitions among federal government regulators, wetland scientists, environmentalists and property owners that have became a source of controversy well into the 1990s. This era illustrates the loss of control by the agricultural monopoly over wetlands policy. Clearly, Era III is characterized as a struggle that is taking place between the monopoly of Era I and the emergent challengers of Era II.

Not until the 1960s and 1970s did Congressional policy significantly preserve and protect wetlands. Unlike previous eras, now several agencies are actively involved with implementing wetlands policy. These include the ACOE, the FWS, and the USDA. Each agency received mandates from the diverse collection of committees and subcommittees that have entered into wetlands policymaking in Congress. Often these committees do not coordinate policymaking activities. The result is an incoherent and conflictual wetlands policy required to be implemented. One example of this struggle for wetlands policymaking was symbolized by victories for major programs by both the agricultural and development interests and the environmental protection interest. For environmentalists, the Wetlands Loan Act of 1961 and Water Bank Act of 1970 reflected the concerns of wetland protection. With the creation of Section 404 in the Clean Water Act in 1972, environmentalists won a significant victory over the policy debate. In addition, federal assistance for wetlands conversion terminated with President Carter's Executive Order 11990 in 1977. That year, the ACOE issued final regulations that expanded their jurisdiction of wetlands to include not just navigable waters as outlined in the 1972 Clean Water Act but also isolated waters. Congress supported the increased jurisdiction for the ACOE in the 1977 amendments for the protection of wetlands. As a result, during the 1980s the conversion rates of wetlands to agricultural lands were at an all time low of approximately 290,000 acres a year (Dahl and Johnson 1991).

Table 3.1 reflects the initial policy victories by the environmental interest groups during the mid-1960s to mid-1970s. However, a tremendous policy conflict began in the mid-1970s and continues in Congress. Today's tone is reflected in the larger number of hearings in the environmental protection venue than previous eras. This point should not be overstated since competition for dominance of wetlands policy is without a clear monopoly even today. Fragmented policymaking is also demonstrated in the number of Congressional bodies conducting hearings in this period. Twenty-one committees and forty-four subcommittees are involved in the debate compared to the four and five committees in Era I and Era II. Of course, the tremendous rise in subcommittees from three in the previous era to forty-four in Era III reflects how specialization in Congress is used to expand committee jurisdiction. This specialization has not shown to be a product of Congressional reforms that occurred in 1974 (King 1994). Instead, it reflects the expansion of jurisdiction of committees not previously in the wetlands policymaking arena. In fact, these "turf wars" by committee were codified in the 1974 reform rather than reforms making any changes on Congressional policymaking (King 1997).

The data also indicates a clear shift has occurred from Era I to Era III. In Era I, sixty-eight percent of the hearings were conducted in agriculturally-related committees with a decline to fourteen percent in Era II. This decline continued in Era III with only nine percent of the hearings being conducted by agriculturally-related committees. This occurs at the same time as there are more interests represented in Congress and more venues available for policy losers. Interestingly, Era I started nonlegislative hearings at a high of sixty-eight percent of all hearings in that period. This number declined to only twenty-nine percent in Era II and never returned in Era III (38%) to Era I levels. This indicates that nonlegislative hearings are not a tool widely used by wetlands agenda setters to alter the wetlands policy definition.

In the mid-1980s, Congress recognized that the national wetlands protection policies and agricultural policies were working at opposing purposes (National Audubon Society 1996). As a result, a swampbuster provision was included in the 1985 Farm Act to reverse Congressional encouragement of wetlands conversion for crop production. Before 1985, farmers who drained wetlands for crops were automatically eligible for agricultural support payments. The provision denied these payments and other agricultural program benefits such as crop insurance, price support payments, disaster payments, storage facilities and loans.

More conflict erupted over these swampbuster provisions that led to the 1996 amendments to the Farm Act that gave the USDA flexibility in enforcing wetlands protection and discretion in the withholding of payments and agricultural benefits. The USDA recaptured being the sole agency in charge of agriculturally-related wetlands issues instead of sharing that domain with the FWS, ACOE or U.S. Environmental Protection Agency (EPA). For the environmental protection advocates, the 1980s included a Tax Reform Act (1986) and the Emergency Wetlands Act (1986) that protected wetlands. It appears that unlike Era I, there is no clear monopoly in Congress that dominates wetlands policymaking. Proponents of agricultural and development issues as well as environmental protection appear to have been effective in venue-shopping for sympathetic committees in Congress in Era III.

In the late 1980s, and early 1990s, the conflict on the science delineating a wetland became part of Congressional hearings. In 1988, the National Wetlands Policy Forum, created to review wetlands policy in the United States, recommended a no net loss of wetlands and a net gain in wetlands as a national policy. This policy was adopted by then

President Bush. The North American Wetland Conservation Act (1989) and the Coastal Restoration Act (1989) created a wetland trust fund to acquire and restore wetlands. An interagency effort among the FWS, the EPA, and the Soil Conservation Service was conducted in 1989 in order to provide a joint and single delineation or identification manual for wetlands. This manual defined in clear terms the criteria for being a wetlands that would be regulated. This allowed competing agencies under the Departments of Agriculture and Interior, along with the EPA, to provide a consensus on implementation of wetlands protection policies. Critics of the 1989 manual claimed that it did not alleviate inconsistencies among the bureaucracies but represented an expansion of regulatory jurisdiction. This manual was thought to be proposing a policy that was too intrusive to property owners and overly protective of wetlands. Former President Bush rescinded the 1989 manual and required the agencies to rework it. As a result, a 1991 manual was issued which reduced protection of the wetlands and was criticized by environmental interest groups as being technically flawed. A policy stalemate resulted between the agriculturists and property owners and the environmental activists, which were often joined by groups who used the wetlands for recreational purposes.

While Table 3.1 reflects the tremendous growth in participation by the environmental protection interests, they do not dominate the policy agenda. For instance, changes were made to the swampbuster provision in the Farm Act of 1990 that concerned many environmentalists. Although this Act established a Wetland Reserve Program to restore wetlands on farmed and prior converted areas, it also allowed for mitigation programs in support of the agricultural and development interests. These mitigation programs allow farmers and others to mitigate a wetland conversion through restoration, enhancement, or creation of a wetland in the same general local watershed. In 1993, President Clinton created the Interagency Working Group on Federal Wetlands Policy to build common ground between the agricultural and development side and the environmental protection side. While this Administration has continued the goal of no net loss established by the Bush Administration, the preference is toward nonregulatory programs such as restoration, mitigation, and public/private partnerships. According to Clinton's Interagency Working Group on Federal Wetlands (1993) approximately fifty-three million acres of prior converted croplands are exempt from federal regulations; lands which were converted to agricultural uses are not subject to wetlands regulations; and the Soil Conservation Service (now called the Natural Conservation Resource Agency) under the USDA is the lead agency for wetlands on agricultural lands under both the Clean Water Act and the Farm Act. The debate continues today with a focus on the agricultural and development side over protection under the Fifth Amendment of the Constitution with "takings" of private property by the government. The environmental protection side continues to express concerns about the use of mitigation banking (Silverstein 1994). In fact, many environmental protection advocates feel that the Clinton Administration really has allowed the conflict from the Bush Administration to continue without any aid in promoting a consistent wetlands policy (Lenetsky, 1994).

Today, the debate also includes more diverse issues regarding the value of wetlands. The role wetlands play as habitat, particularly for endangered species, is of interest to wetlands policymakers. Research is just beginning to tie wetlands to these critical habitats. One study finds that significant numbers of threatened and endangered species are associated with wetlands habitats (Boylan and MacLean 1997). Approximately forty-six percent of all endangered and threatened species are wetlands associated or dependent on the habitat (Boylan and MacLean 1997; 14). The authors find that the relationship of habitat degradation to wetlands loss and correlation to species endangerment is still relatively unknown in the scientific community.

This conflict is reflected in the witness testimony of this era. Many more witnesses testified in this era (1446) than in the previous era. In comparing which committees most witnesses testified before, a significant change has occurred. Era I had a majority of witnesses testifying before agriculturally-related committees (52%). In Era II, the majority of witnesses testified before the Merchant Marine and Fisheries Committee (42%). In Era III, the Merchant Marines and Fisheries (21%)

shares an equal number of witnesses with Environment and Public Works committees (20%). Because of the great diversity of venues in Congress, no subcommittee has greater than eight percent of the witnesses with the majority of subcommittees having less than five percent of all the witnesses for this era.

A major shift has occurred not only in venues but also in witnesses who testify. Unlike the previous eras, federal elected officials (7%) are less involved in testifying before Congress. Federal agencies (25%) and interest groups (20%) are the largest groups represented. This indicates that elected officials who testify in hearings are no longer a major policy influencer for wetlands agenda in Congress. Instead, the conflict among the FWS, USDA, and EPA has been taken into the Congressional testimony along with the interest group participation. Clearly, elected officials played a role in the agricultural dominance period and stability of Era I and other elected officials helped venue-shoppers to redefine the wetlands agendas in Era II. While witnesses (46%) continue to geographically represent the Washington, D.C. area, many more groups are represented than in previous eras with more diverse issues. For example, when evaluating key issues in witness testimony, topics include concerns about sport/leisure, property rights, agriculture, and migratory animals. The most prevalent key issue was the protection of wetlands (28%). Like the previous eras, no single witness dominates in testifying before Congress on wetlands policy. In fact, the same witness testifies before Congress less than five percent of the time.

Testing the Relationships among Wetland Policymaking Venues, Witnesses and Hearings from 1789–1999

Using the data from all three eras, the relationship among witnesses, hearings, and venues is tested based on tone. Congressional policy dynamics have a significant role in helping promote a certain type of wetlands policy. Congressional committees engage in this process through three main routes. First, committees influence the definition or redefinition of a policy by conducting hearings with a certain tone. That is, instead of considering issues of the opposing tone, committees only conduct hearings reflecting in their tone. The topic being considered usually is not neutral and the subject of the hearing not open to consider a wide range of alternatives. Second, witnesses who testify before a committee can either reflect the tone of the committee or an opposing tone. Committees concerned about maintaining or expanding their jurisdiction have an incentive to select witnesses that have a corresponding tone. These two dynamics actually reinforce each other by satisfying the needs of policy advocates who need to venue-shop because their interests are not being considered by the status quo in Congress and by committees seeking to expand their jurisdiction. To investigate if these Congressional dynamics exist in wetlands policy-making, a correlation should occur between tone of the hearings and committees, and between tone of the committees and the witnesses. To investigate for venue-shopping and Congressional expansion, two tests were conducted using Chi-Square and correlation coefficients of Gamma and Kendall's Tau-b. In addition to the two categories of tone, a third category of other was used for hearings and committees that were neutral in tone or uncodeable.

The first test was a crosstabulation between the tone of the committee's jurisdiction and the tone of the hearing. Table 3.2 shows that the committees tend to conduct hearings that reflect their defini-tion of the policy (Chi-Square = 370; $p < .00001$; Gamma = .96 and Kendall's Tau-b = .86 $p < .00001$). Environmental committees tend to conduct hearings having an environmental tone versus other interests (92.2%). Likewise, agricultural and development committees almost always conduct hearings with a tone reflective of their interests (95%). Thus, this data shows that tone of hearings and committee jurisdictions are excellent ways to understand how Congress helps define wetlands policy. Committees are using hearings to garner support for the views they already hold and establish their jurisdiction over a certain aspect of wetlands policymaking.

While these results are similar to the findings of Jones et al. what is different in wetlands hearings is who testifies. From the analysis of the three eras of wetlands policy, no single witness or representative group had a large monopoly over another testifying before Congress. Likewise, Table 3.3 shows that committees conducting hearings on

Table 3.2 Relationship between Tone and Venue
in Congressional Hearings on Wetlands

Venue	Tone of Hearings			
(Committees)	Agricultural and Development Tone	Environmental Protection Tone	Neutral or Uncodeable Tone	TOTALS (N)
Agricultural and Development Tone	95% (76)	5% (4)	0% (0)	100% (80)
Environmental Protection Tone	7.8% (9)	92.2% (106)	0% (0)	100% (115)
Neutral or Uncodeable Tone	5.6% (3)	13% (7)	81.5% (44)	100% (54)
Totals	35.3% (88)	47% (117)	17.7% (44)	100% (249)

Gamma = 0.96, Kendall's Tau-b = 0.86, Chi-Square = 370 (p < 0.00001)

wetlands policies hear from interests unlike their own (Gamma = .37 and Kendall's Tau-b = .21, p < .00001). Environmental committees have more of a slight tendency to hear from environmental witnesses (56.6%) than agricultural related committees (49%). Agricultural committees were more likely to consider an opposite viewpoint from the environmental protection tone (41.8%) than were environmental committees willing to hear from an opposing agricultural tone (25.9%).

Overall, nonlegislative hearings were less useful as a tool for Congressional committees to redefine wetlands policy in order to increase their jurisdiction. And while committees tend to conduct hearings with a similar tone to help define or redefine wetlands policy, these same committees do tend to hear between twenty-five and fifty percent of the time from witnesses with a different tone. Thus, venue-shopping for wetlands committees may allow some accessibility with agricultural committees that conduct hearings.

The third Congressional dynamic that assists with policy changes is the relationship between Congressional attention and venues. As Congressional attention on wetlands increases, the number of venues claiming jurisdiction also should increase. Congressional attention to

**Table 3.3 Relationship between Venues and Witnesses
in Congressional Hearings on Wetlands**

Venue	Witness Tone			
(Committees)	Agricultural and Development Tone	Environmental Protection Tone	Neutral or Uncodeable Tone	Totals (N)
Agricultural and Development Tone	49% (246)	41.8% (210)	9.2% (46)	100% (502)
Environmental Protection Tone	25.9% (264)	56.6% (642)	11% (112)	100% (1018)
Neutral or Uncodeable Tone	13.5% (13)	69.8% (67)	16.7% (16)	100% (96)
Totals	32.4% (523)	56.9% (919)	10.8% (174)	100% (1616)

Gamma = 0.37, Kendall's Tau-b = 0.21, Chi-Square = 100 (p < 0.00001)

wetlands attracts new venues to capture jurisdictions that are unclaimed by others. An ordinary least squares regression, used to test for the how much Congressional attention is paid to wetlands, explains the growth of venues claiming wetlands jurisdiction. Equation 3.1 shows the ordinary least squares equation used in this analysis.

Congressional attention, measured as the number of hearings on wetlands per year (independent variable), results in increases to the number of available wetlands venues as measured by the number of

---

Equation 3.1 Wetlands Congressional Attention and the Role of Committees

$$Y_t = \beta_1 X_t + \beta_2 Y_{t-1} + e_t$$

Y_t = Number of Wetlands hearings conducted each year on the environmental issue (dependent variable).

X_t = Congressional Committees holding Hearings on wetlands each year (independent variable).

Y_{t-1} = Number of Hearings conducted previous year on wetlands (leg variable).

e_t = residual unexplained by model.

subcommittees (dependent variable). Because of the large activity in subcommittees starting in Era II, they were used as indicators of attention rather than committees. It is not useful to include years with almost no variance in the number in the hearings and committees, which occurred in Era I.

In addition, there is some debate on the role political parties play in venues and hearings in Congress. One would think perhaps majority party plays in Congress would correlate with the tone of hearings and venues. Several authors suggest that historically it is difficult to make inferences about the role the majority political party in Congress. One exception to this occurs in the 1980. From 1952 to 1964, there is little change in partisanship in Congress (Aldrich 1995). From 1965 to 1977, during the early part of Era III, Democrats from the South deflected from the party into a Conservative Coalition with Republicans. Later in the 1980s, there was a revival of partisan voting patterns with the return of southern Democrats. Rohde (1991) finds evidence that political parties in the House of Representatives grew stronger in the 1980s. However, Krehbiel (1993) shows that this reflects no more than different preferences of members of Congress that became more closely associated to partisan affiliations in 1980s. Thus, party affiliation indicated what preferences were brought to Congress rather than majority party in Congress impacting the fragmentation of venues and issues. Therefore, party seems to be defined by the office seeker rather than an independent variable explaining Congressional attention to wetlands. To test for any impact party may have on the number of venues, party was entered as a variable in the regression equation and tested for in all three environmental cases. Like the other studies, it appears that party, as measured by either the number of democrats or as a dichotomous variable of majority party in both chambers, showed no significant relationship.

The regression results in Table 3.4 indicate that as Congressional attention increases, Congressional venues proliferate to accommodate venue-shoppers and committees seeking to expand their jurisdiction (adjusted R Square = .85; Beta = .62; p < .0001). In fact, the results show that Congressional attention has a large impact on the expansion

**Table 3.4 Relationship between Congressional Venues and Hearings
in Wetlands Policy from 1789 to 1999**

Variable	B	Standard Error	Beta Coefficient	Significance
Intercept	0.07	0.07	0.31	
Wetland Hearings	0.32	0.03	0.62	< 0.0001
F Statistic				< 0.0001
Adjusted R-Square	= 0.85			
Durbin Watson*	= 2.11			
N	= 153			

*Indicates no autocorrelation problem. The dependent variable is annual number of subcommittees holding hearings on wetlands.

of venues. A nice triangulation of results is demonstrated when Table 3.1 compares the eras with the Chi-Square tests and regression analysis and all are taken into consideration. All results point to a decline of a significant monopoly that once had centralized venues and attention around a tone based on agriculture and development.

Wetlands as a Dominance Issue Model

Two important conclusions are drawn from this study on wetlands policymaking in Congress. First, major redefinitions of wetlands in Congress have a significant impact on management and implementation of wetlands policy. This redefinition process has caused significant conflict in Congress that continues today. It also has impacted the natural resource. Second, predictions are made for the potential of this issue to transform into another model of issue definition. Both of these conclusions drawn from the study indicate that understanding the issue definition process is critical to the policy development and agenda setting involved with wetlands policymaking. It is clear that without taking tone into account in understanding agenda setting, the agenda for wetlands policy could not be explained. Redefinition of the concept of wetlands and the change in tone both were key factors in the existence of wetlands as a natural resource today.

Redefining Wetlands

Congressional wetlands policymaking has gone through significant policy changes. These changes can be classified into three eras that reflect a dominance issue model. Until recently, wetlands policymaking in Congress has focused on issues defined primarily around agricultural and development issues resulting in the baseline tone of wetlands definition by the agricultural and development interests. Evidence for this conclusion comes from the examination of committees and witnesses who testify before them. In Era I (1789–1945), wetlands were defined as lands that needed to be improved for human needs. The evidence from examining Congressional hearings clearly shows the dominance of agricultural and development concerns defining policy-making. Moreover, there was a small monopoly that was tightly held in Congress. For instance, a small number of committees were involved in holding hearings. Witness testimony reflected these narrow concerns of interest to the committees.

This predominant view of wetlands had devastating impact as demonstrated by the many studies documenting wetlands loss in this era (Mitsch and Gosselink 1993:43). In Era II, from 1946 to 1966, the definition of wetlands changes in Congress. This redefinition process slowly emerges in Era II through evidence of new committees and witnesses participating in the hearing process. Interestingly, it is not enough to look only at nonlegislative hearings to find this transition. Other studies have shown that nonlegislative hearings can aid committees in staking out new claims of jurisdiction (King 1994). However, many committees acquired legislative control over unclaimed territory of environmental aspects to wetlands management. The data very clearly shows new committees holding hearings that were not previously involved in the policymaking. These new participants were not concerned with the old definition of wetlands as a land development issue for agricultural and human needs. Instead, these new venues and witnesses reflected a redefined wetlands policy that focused on environmental protection. There are several socioeconomic trends that add to this redefinition and it should not be interpreted that redefinition takes place in isolation of societal concerns. Just the contrary, Congressional hearings and committees often reflect the combination of several socioeconomic impacts. The tremendous loss of wetlands to a growing population moving to the suburbs, and the sophistication of wetlands science aided in the larger picture of wetlands redefinition demonstrated in Congressional policymaking.

From 1966 to 1999, these different definitions of wetlands have both found venues in Congress. Though the dominance of agricultural and development issues no longer has a monopoly in Congress, it has not been eliminated. Instead, a highly conflictual debate is ongoing between these two approaches to wetlands policy. Because the dominant venue from Era I did not encompass environmental concerns, new committees aggressively sought unclaimed territory. With the advent of environmental management, environmental protection interests also found new receptive venues that were once closed to their concerns.

No one individual witness has prevailed in testifying before Congress. Another interesting finding is that few women testify before Congress even though we have seen a growth in the number of females in senior level positions in organizations. As a group, federal agencies and elected officials tend to testify more than others. Currently, federal agencies and interest groups are the most frequent witnesses testifying before Congress on wetlands. Nonetheless, a great diversity of groups and many different participants have been represented.

Future of Wetlands

It is difficult to predict exactly what course wetlands policy will take in the future. To date, this issue has demonstrated a pattern that typifies the dominance issue model. Currently, the issue is undergoing turmoil in issue definition that has caused significant conflict in implementation by federal agencies. Since the agencies such as the ACOE, USDA, FWS and U.S. EPA are well established in the area of wetlands implementation as well as the Congressional committees jurisdictions, perhaps wetlands Congressional policymaking can evolve into a more bounded issue. An issue reflecting one pattern of issue definition does not preclude it from transforming into another model. The difficulty with wetlands is that the issue has been very conflictual at times with even presidential involvement in the definition of a wetland. It would be difficult for this transformation into a bounded issue model to occur under these conditions. What is interesting is that when an agricultural tone for wetlands policy shifted to an environmental one, the path the wetlands agenda took was one of conflict. Other paths or options could have been taken like a bounded issue model or shared coexistence on the agenda. Instead, wetlands policy continues today as a conflictual, divisive policy.

Clearly, wetlands policymaking in Congress is in the midst of a new phase. What is unclear is the actual path the issue will take. It is very possible that the issue of wetlands will remain on a dominance pattern with continued conflict and periods of stability. As the new century begins, it is uncertain if very opposing sides of the wetlands issue can actually coexist. Congress continues to be a battleground for defining a wetland for both environmentalists and developers. The next chapter examines the advantages of having an issue take a different path than one displayed here in wetlands. The advantages of having wetlands transform into more of a bounded issue model as exemplified by the Great Lakes issue holds some benefits for the natural resource. The next chapter looks at a different path and how it can hold benefits for the natural resource.

CHAPTER 4

Bounded Issue Model—
The Case of the Great Lakes

The United States Great Lakes (referred to as the Great Lakes or Lakes) are one of the most important natural resources in the country, if not the world. Not only do the Great Lakes span over 750 miles across eight states and two Canadian provinces, they also contain eighteen percent of the world's and ninety-five percent of the United States' fresh water supply (U.S. EPA 1995), with only the polar ice caps containing more fresh water. The Great Lakes cover about one-third of the border between Canada and the United States and themselves border Minnesota, Wisconsin, Illinois, Indiana, Michigan, Ohio, Pennsylvania, and New York as well as Ontario and Quebec. The nearly 300,000 square mile drainage basin is home to one-tenth the population of United States and one-quarter that of Canada (U.S. EPA1995). Nearly twenty-five percent of the total Canadian agricultural production and seven percent of the American production are located in the basin. The Lakes contain 5,500 cubic miles of water covering a total area of 94,000 square miles. Currently, the basin provides eleven percent of total employment and fifteen percent of manufacturing jobs in Canada and the United States (Gleick 1993). The consequences of industry, commerce and navigation became obvious by the twentieth century. Entire food webs were altered with the pollutants, over-fishing, and introduction of exotic species like zebra mussels or sea lamprey. Beaches today continue to periodically be closed because of bacteria from human waste.

Results show that three major eras occur in Congressional policymaking on the Great Lakes based on the tone of hearings conducted. This is based on the percentage of hearings in each tone. During Era I that includes the years from 1789 to 1965, Congressional policymaking is characterized by a focus or emphasis of tone on industry, commerce and navigation concerns. It was not until Era II (1966–1983) that concerns about the Great Lakes shifted focus to an environmental protection and health tone. This new tone challenged the prior dominant interests of Era I, and to a significant extent, almost replaced it. Since 1984, however, a new Era has emerged in Congressional policymaking on the Great Lakes. Unlike the previous eras, Era III (1984–1999) reflects an agenda sharing of both environmental and industrial issues. This agenda sharing is critical for natural resources that are viewed as part of an ecosystem which requires sustainable development not exhaustible use as in the case with wetlands. This agenda sharing is the most beneficial for the natural resource since it provides protection and use of the resource without its overconsumption. The current Congressional policymaking indicates that there is a lack of dominance in the Great Lakes Congressional policymaking. This chapter addresses how the Great Lakes policymaking in Congress has changed over time. Then, the chapter reports findings using statistical tests on tone of witnesses, hearings and venues along with a regression analysis lending evidence to how these eras took place in Congress. The conclusion explains how the Great Lakes policymaking resembles a bounded issue model that continues today and the implications of that model on the natural resource.

**Relationship of Congressional Hearings
and Committee Dominance in Great Lakes Policymaking**

Congressional policymaking on the Great Lakes in the United States has had significant changes over time. There are three distinct eras based on Congressional activity. Using the CIS Abstracts on CD-ROM for Congressional hearings from 1789–1999, a search was performed on the keyword the Great Lakes for the entire text of the hearing abstracts, testimony summaries and witnesses. This search yielded

379 hearings with 3513 witnesses, which reflect the complete set of Congressional hearings on the Great Lakes in the United States.

Hearings, committees and witnesses were coded into detailed categories on all aspects of the Great Lakes. From this detailed coding, two major themes emerged for analysis. Hearings, committees and witnesses tended to focus on two major key issues which are: (1) navigation, industry, and commerce, or (2) environment and health. A third category was used for neutral or uncodeable issues. For instance, a hearing conducted on improvement to the St. Lawrence Seaway was coded as the first category on industry, commerce, or navigation. Likewise, a hearing conducted on concerns about water quality, recreation, public drinking water or fish consumption was all coded as the second category on environment or health. The third category comprised hearings such as appropriations for funding the Great Lakes research, which were considered neutral depending on the testimony. Often committees and witnesses combine issues of environment and health when testifying about natural resources, particularly the Great Lakes. Likewise, industry, commerce, and navigation are naturally interrelated topics. This combination of key issues into two tones was the most appropriate coding scheme for analyzing hearings, committees, and witnesses.

These categories are useful for the analysis because they reflect significant differences in Congressional policymaking of the Great Lakes. Every committee, subcommittee, and witness was given a specific code to identify monopolies of jurisdictional control and dominance by tracking their participation over time. Figure 4.1 shows the number of annual Congressional hearings conducted on the Great Lakes policy. In general, Figure 4.1 shows a tremendous increase in overall hearings conducted on the Great Lakes in the last several decades. Besides tone, this information was coded for supplemental data information such as gender of witness and organizations represented. Individual codes for each witness were used to track if witnesses testified multiple times indicating a reliance on certain interests by Congress.

Unlike the wetlands issue described in Chapter Three, the Great Lakes hearings have continued thoughout time. This resource was

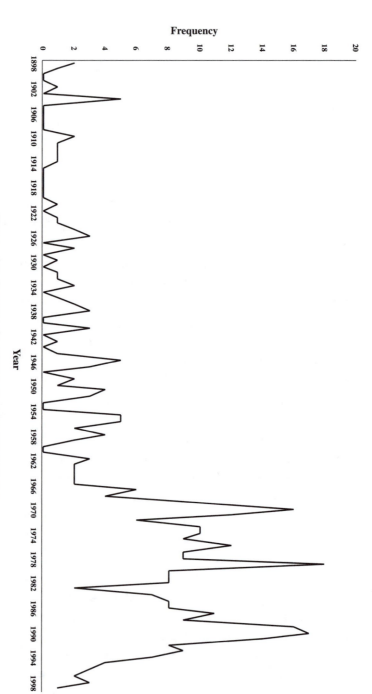

Figure 4.1 Congressional Hearings on Great Lakes

always considered an important one for any definitional tone describing it. This is reflected in the periodic yet somewhat consistent number of hearings over time without long periods of no Congressional activity. Again, this is unlike the wetlands profile of hearings.

In addition, hearings are coded as legislative or nonlegislative. Hearings that consider bill-referrals are defined as legislative. Generally, committees tend to use nonlegislative hearings to encroach on the juris-dictional control of another committee, which would indicate a chal-lenge to a committee's monopoly. Nonlegislative hearings are particu-larly important because few restrictions exist on the subject of hearings.

Congress made the Great Lakes policy using forty-eight commit-tees and seventy-two subcommittees. Two hearings were held in 1898 and one in 1899, which are not shown on Figure 4.1. The number of witnesses who testify on the Great Lakes annually is displayed in Figure 4.2. Like the hearing profile in Figure 4.1, it appears Congress has a continual or enduring interest in the Great Lakes since there is no long period of hearing inactivity displayed in the wetlands profile. No hear-ings were conducted before 1898 on the Great Lakes. The next section on results describes in detail the three eras in Congressional hearings on the Great Lakes.

Results

Three Eras of Congressional Policymaking on the Great Lakes

Of the 379 total hearings conducted, most were in the House (70%) and considered formal legislation (77%). For the total set of hearings, there was approximately equal percentage of them that were coded into the categories of key issues: navigation, industry, or commerce (36%), environment or health (31%), and neutral or uncodeable (32%). When this data is disaggregated, the three eras emerge. Close to three-fourths of all the hearings (74%) were conducted in the last twenty-seven years from 1967 to 1999. In fact, Figure 4.1 shows that beginning in the mid-1960s an unprecedented amount of activity begins in Congress regarding the number of hearings conducted each year. With the exception of 1984, when there are only a few hearings

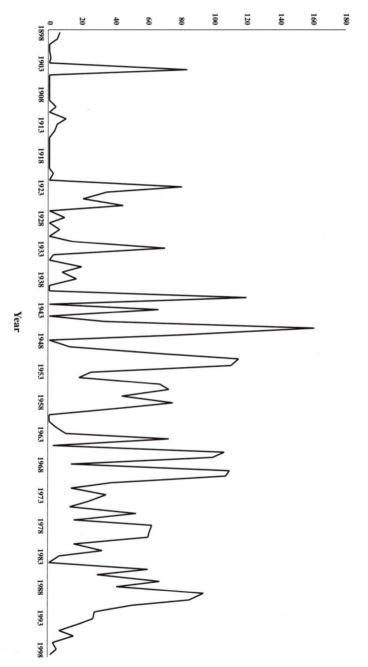

Figure 4.2 Witnesses who Testify on Great Lakes

conducted on the Great Lakes, the overall average frequency of hearings has remained significantly higher than prior to the mid-1960s (with some decline at the end of the 1990s). Therefore, the new dominant tone of environment and health replaces the industry, commerce and navigation tone of Era I but it is not nearly as strong and enduring as the old dominant tone. Finally, Era III (1984–1999) reflects a lack of dominance by either tone. For each era, Table 4.1 breaks down the venue and witness information presented in the following section.

Era I—Navigation, Industry, and Commerce (1789–1965)

From 1789 to 1965, most of the Congressional attention on the Great Lakes was focused on concerns about using them for navigation, industry, or commerce (Ashworth 1987). The Great Lakes basin before the 1600s had a population of aboriginal people estimated between 60,000 and 117,000 (U.S. EPA 1995). In the sixteenth century when Europeans began their search for a passage to the orient, these people moved westward as the Lakes were discovered by the Europeans. The Great Lakes were originally conquered and viewed as a navigation resource. By the early 1600s, the French had discovered the forest around the St. Lawrence area and used beaver fur for trade. Samuel de Champlain and Etienne Brule, his scout, were the first explorers to reach the Lakes. By 1670, the first French forts were built around the basin to protect beaver fur trading near the Straits of Mackinac. In 1673, the first fort on the lower Lakes was built at Kingston, Ontario.

The British gained control of Lake Ontario in 1727 and maintained control of the Lakes during the American Revolution. The Lakes were the official boundary between the new United States and the British colonies. Between 1792 and 1800 the population of Canada increased from 20,000 to 60,000 people. With the War of 1812, which lasted two years, the Americans secured ownership of the American side of the Lakes while the British maintained their upper Canadian portion.

As settlers moved West, the Great Lakes were a fast and cheap transportation route linking the rapidly expanding east coast to the Midwest. Therefore, the building of ports and harbors, as well as the management of transit on the Lakes such as building lighthouses or

Table 4.1 Congressional Great Lakes Policymaking from 1789–1999

	Era I (1789–1965) Navigation, Industry, Commerce Dominance	Era II (1966–1983) Shift to Environment and Health Dominance	Era III (1984–1999) Agenda Sharing
A. Hearing Information			
Number of Hearings	93	159	127
Number of Committees	14	17	17
Number of Subcommittees	11	30	31
Percentage of Hearings Held by Each Committee			
Merchant Marines and Fisheries Committee	22%	18%	30%
Public Works	17%	0	4%
Rivers and Harbors	22%	0	0
All other Committees	< 5%	< 8%	< 13%
Percentage of Hearings Held by Each Subcommittees	All < 5%	All < 10%	All < 10%
Type of Hearing			
Legislative	65%	86%	76%
Nonlegislative	36%	14%	24%
Type of Committees			
Commerce	14%	0%	14%
Public Works	30%	9%	9%
Navigation	28%	4%	2%
Environmental	7%	33%	32%
B. Witness Information			
Number of Witnesses	2098	858	557
Gender of Witnesses			
Male	1795	795	473
Female	10	28	62
Uncodeable	276	35	22
Percentage of Witnesses that testify before each Committee			
Merchant Marines and Fisheries	7%	22%	33%
Public Works	28%	14%	28%
Foreign Affairs	32%	2%	< 1%
Rivers and Harbors	17%	0	0
Percentage of witnesses testifying before each subcommittee	All < 8%	All < 9%	Water Resources 17% Others less than 9%
Percentage of Selected Witnesses			
Trade Associations	23%	10%	8%
State and Local Agencies	16%	10%	13%
Federal Elected Officials	15%	11%	7%
Private Sector	11%	5%	4%
Interest Groups	2%	7%	9%
Federal Government	9%	21%	25%
Specific Witnesses Testifying			
Chambers of Commerce	5%	< 1%	< 1%
Environmental Protection Agency	NA	3%	6%
Department of Defense (Environmental Quality Division)	NA	5%	< 1%

Percentages are calculated based on each era. Only major percentages are presented. All categories for each variable are not included because they reflected a smaller percentage.

regulation of maritime commerce, were the early definitions for the Lakes. Industry such as timber, agriculture, and mining helped cultivate the growth of the urban centers around the Great Lakes in the 1800s. By 1830s, commercial logging began in upper Canada, Michigan, Minnesota, and Wisconsin. The degradation of the Great Lakes began early in the settlement history of the basin with the first paper mill built in 1860 on the Welland Canal. The abundant forests surrounding the basin support seventy-three plants for pulp and paper industry. These plants discharged large amounts of chlorine and mercury compounds into the air and water.

The agricultural uses of the fertile basin also deposited large amounts of fertilizers and pesticides from runoff. Many of the wetlands and forests of the Great Lakes were converted into agricultural lands, along with residential and industrial uses. In the mid-1800s, the greatest attraction for immigrants to the basin was the availability of cheap, fertile, agricultural lands. The production of grains, dairy, and meat grew as population exploded in growth for the basin.

This exploding population of the basin in the middle decades of this century placed a strain on the ability of the Lakes to handle the increasing sewage it received. As towns and cities grew, inadequate sewage treatment meant that the Great Lakes became a recipient of large amounts of pollutants from both industry and local households. By the 1950s, the populations of bald eagles were declining along the shores of Lakes Erie and Ontario. By the 1980s, minimal numbers of nests were found along the shores of Lakes Superior and Huron. The declining animal populations were some of the early signs of problems with the Lakes. Small mammals throughout the basin that fed on fish from the Lakes were bearing dead young (The Great Lakes Water Quality Agreement Backgrounder 2000). By the middle of the twentieth century, traditional uses of the Lakes were lost due to human misuse of the Lakes. For instance, sport fisheries closed, swimming was restricted, and drinking water had to be purified before consumption.

In 1909, Canada and the United States signed a landmark agreement. In the Boundary Waters Treaty, the two nations agreed to not pollute the Great Lakes. This treaty also established the International

Joint Commission between the U.S. and Canada. Even with this treaty, the Lakes were contaminated with pollutants and exposed to exotic species from their industrial, commercial, and navigational uses. By 1970, pollution in the Great Lakes had reached almost disastrous levels (The Great Lakes Water Quality Agreement Backgrounder 2000). When the early French explorers, like Etienne Brule, discovered Lake Huron in 1612, Lake Ontario in 1615, and Lake Superior in 1622, the basin was very different than it is today. When Jean Nicollet, a French explorer, discovered Lake Michigan in 1634, and Louis Joliet found Lake Erie in 1669, the Lakes were pristine in water quality and the ecosystem balance was naturally preserved.

To accommodate this population growth, the basin begins to develop rapidly with a series of canals. By 1825, the Erie Canal, stretching 364 miles from Albany to Buffalo, opens bringing the largest numbers of settlers to the Great Lakes basin and shipping freight east ward. By 1829, the Welland Canal, about twenty-seven miles, opens to bypass the Niagara Falls and to connect Lakes Erie and Ontario. Already in 1850, the Lakes support the populations of Michigan, Wisconsin, and upper Canada which total 1.2 million people.

Later in the mid-century, railroads tend to replace much of the commerce shipped on the Lakes. The 1959 completion of the St. Lawrence Seaway allows large ocean vessels to enter the Lakes bringing exotic species from the Baltic region in Europe to disrupt the aquatic ecosystem. Today, the main commodities shipped on the Lakes are iron ore, coal, and grain.

Commercial fishing begins around the 1820s with the largest fish harvests recorded in 1889 and 1899 at 147 million pounds (U.S. EPA 1995). Since the 1950s, the average annual catches have been about 110 million pounds. The overall value of fisheries has declined due to larger more desirable fish giving way to smaller, less-valued species. Today, lake trout, sturgeon, and lake herring survive in reduced numbers when they once were abundant. These species have been replaced by human introduction of smelt, alewife, splake, and pacific salmon.

The stress on the ecosystem is first indicated in the 1854 cholera epidemic in Chicago that is caused from the sewage contamination of drinking water from the Lakes in which five percent of the population dies. As early as 1870, Hamilton, Ontario can no longer draw drinking water from the Lakes because of contamination. Again, Chicago in 1891 is plagued with a typhoid epidemic, which reaches 124 deaths per 100,000 people, reminding people of the sewage contamination of drinking water. This causes Chicago in the late1890s to reverse the flow of the Chicago River away from Lake Michigan. This project by the ACOE literally moves more dirt than the Panama Canal.

Population growth booms in the years 1900 through 1965 in this area of the country. By as early as 1900, the basin population reaches 11.5 million people. In just ten short years, the basin reaches a population of 12.5 million. The fastest growth takes place in this time period. By 1930, the basin's population reaches 22.7 million, an increase of thirty-nine percent in twenty years. And, by 1950 the basin population is 22.7 million, which represents an increase of thirty percent in twenty years. At the end of 1959, the St. Lawrence Seaway opens, and the degradation is more than clear to people living in the basin.

From this need for transit of goods and people, the focus of the Great Lakes policy shifted to maintaining an economic resource through building and management locks, seaways, and canals. Because the Great Lakes were such a vital resource, even during times of scarcity of resources such as World War I and World War II, Congress continued to appropriate funding into maintaining the navigation and industrial uses of the Great Lakes, particularly the St. Lawrence Seaway. Table 4.1 shows that compared to the other eras, only ninety-three hearings were conducted from 1789 to 1965. In addition, only fourteen committees and eleven subcommittees were responsible for the Great Lakes agenda. During this dominant era of industry, commerce, and navigation, policymaking in Congress was nonconflictual with no substantial challenges to the status quo. The vast majority of committees conducting hearings came from commerce, public works, and navigation. Only seven percent of the committees involved in Congressional policymaking on the Great Lakes during Era I were concerned with the environment. During this first era, the Great Lakes were to be conquered and used for commerce, industry, and navigation.

The primary Congressional committee that has conducted hearings on the Great Lakes has been the Merchant Marines and Fisheries Committee. Over time, this committee has been able to adapt to the changes in tone for the Great Lakes. This adaptation point is key to have a bounded issue model. For an issue to take a bounded model path, the Congressional participants must evolve to include new definitions of the issue. Since the Merchant Marines and Fisheries Committee started out as more concerned with navigation and maritime commerce, it is critical for it to evolve in agenda focus as the definition of the Great Lakes changes over time. This committee in Era I was primarily focused on the industrial concerns of fisheries and other commercial interests on the Great Lakes. This focus changed in Eras II and III to the Committees being interested in environmental and health issues. Overall, this committee has had an enduring impact on the Great Lakes policy, more than all other committees who conduct hearings much less frequently. This committee, through all three eras, has maintained between eighteen and thirty percent of all the hearings. This is the largest percentage in each era. The Rivers and Harbors Committee (22%) along with the Merchant Marines and Fisheries Committee (22%) reflect the major committees holding hearings. It is not surprising that these committees, along with the Public Works Committee (17%), are the agenda setters during this period of development of the Lakes.

Most hearings during this era were nonlegislative. It was not until Era II and III that there is a great expansion of hearings with the majority of the hearings conducted being legislative. Again, like wetlands, this result is contrary to the Talbert et al. findings that claim nonlegislative hearings are used to increase jurisdiction of committees. In fact, just the opposite occurs in the Great Lakes policymaking. Committees conducted more hearings that were nonlegislative during periods of policy dominance than during periods of redefinition. Legislative hearings dominate as the tool for agenda setting in the Great Lakes policy throughout all the eras.

Witnesses who testify in Era I tend to reflect concerns of trade associations (23%), the public works agencies of state and local governments (16%), and concerns for industry and commerce by private sector interests (11%). Federal elected officials (15%) are also involved since the Great Lakes have such a tremendous impact on industry and inter-state commerce. Interest groups (2%) and the federal government (9%) are minor agenda setters during this period, yet later become more involved in Eras II and III. It was not until Era II and Era III that the concerns about the environment, as most prominently defended by the Federal government, become a major tone in witness testimony.

Era II—The Shift to Environment and Health Dominance (1966–1983)

By 1965, people around the Great Lakes basin were seeing signs of the deterioration of the Lakes. In 1965, there are reports of reproductive failure in Michigan and Ontario by ranch minks that fed on the Great Lakes fish. In 1967, bald eagle breeding declines, and in 1969 the Cuyahoga River in Ohio literally catches fire. Clearly, the Lakes were no longer the natural resource to be relied upon from the earlier era. The damage and overuse of the Lakes had taken its toll on the ecosystem.

Beginning in 1965, a dramatic shift in tone regarding the Great Lakes occurs in Congress. A total of 159 hearings are conducted during this era with 17 committees, 30 subcommittees, and 858 witnesses. Not surprisingly, this corresponds with the larger environmental movement taking place in the United States. The Great Lakes begin to be redefined as a natural resource needed for safe drinking water and a rare fresh water supply in the world. However, there are several unique factors about this new dominance of environment and health that replaces the old tone of industry, commerce and navigation. First, this new domi-nance was accomplished through new committees entering into the Great Lakes policymaking by conducting hearings with an environ-mental tone. Data in Table 4.1 reports that approximately thirteen to seventeen committees were involved over time with the Great Lakes policymaking. Several new committees entered into the jurisdiction of the Great Lakes policymaking in Era II and maintained their roles into Era III. Venue-shopping takes on a different meaning here. In other words, not only can interests look for existing committees seeking to expand jurisdictions, but also along with the environmental movement, can help create new venues in Congress sympathetic to their interests.

In addition, the increase of subcommittees from eleven in Era I, to thirty and thirty-one respectively, in Eras II and III reflect the jurisdiction expansion taking place in Congress.

Second, committees did not accomplish this shift in dominance expanding their jurisdiction through only nonlegislative hearings. Committees and subcommittees held legislative hearings the vast majority of the time (86%) to implement this new dominance. This is very unusual compared to other areas such as health care where committees heavily relied on nonlegislative hearings for jurisdictional control and expansion. In Era I, commerce, public works and navigation committees collectively controlled seventy-two percent of the types of committees. By Era II, this dramatically changed to thirteen percent. Now in Era III, the largest amounts of hearings are conducted by the environmental committees (33%). The River and Harbors Committee along with the Public Works Committee from Era I are essentially not agenda setters in Era II. Finally, not only did the percentage of committees that dominated Era I to Era II change from industry, commerce, and navigation (72%) to environmental (33%), the witnesses from whom the committees heard also changed. The Federal government was the largest (21%) represented witness group that testified in defense of the environment in Era II. Recall in Era I, the Federal government only garnished nine percent of the witnesses who testified. Similar to wetlands, during periods of environmental tone, the Federal government plays a key role in Congressional testimony. Trade associations (10%), state and local agencies (10%), the private sector (5%), and federal elected officials (11%) all play reduced roles from Era I in testifying before Congress. Interestingly, during an era of tremendous environmental tone, interest groups overall only testified seven percent of the time during Era II, which is very different than in the environmental era in the wetlands case in Chapter Three. While the environmental movement is clearly reflected in Era II, another major change takes place in the Great Lakes policymaking in Congress beginning in 1984 with the start of Era III.

Some important policies are established in this era for the Great

Lakes that reflect the concern for environment and health aspects which continue in Era III. For instance, in 1972, Canada and the United States sign the Great Lakes Water Quality Agreement in recognition of the urgent need to improve environmental conditions in the Lakes. The commitment focused on restoring and enhancing the water quality, which included reducing the discharge of pollutants toxic to human, animal, or aquatic life. In addition, numerical targets were used for the reduction of phosphorus to Lakes Erie and Ontario. It was determined in the early 1970s, that Lake Erie was "dead" from the excessive amount of phosphorus from detergents, which was actually causing eutrophication of the lake. In addition, the upper portions of the St. Lawrence River were seriously damaged by chemical pollutants that caused injury to both humans and property in both Canada and the U.S. Later in 1978, the agreement was updated to introduce the concept of ecosystem management that highlights the interconnectedness of all components of the Lakes and the need for an integrated approach in dealing with human health and environmental quality. This agreement amendment also established the elimination of toxic substances and established a list of toxic chemicals for priority action. It is this ecosystem approach, originally advocated by interests from the environmental and health tone, that blossoms into a bounded issue model in Era III. Basically, like wetlands, the Great Lakes agenda has followed the path of a tone set first for use of the resource in Era I with a shift to an environmental tone in Era II. Granted these shifts start in different time frames. For wetlands, the environmental Era II begins in 1946 and continues to 1965, whereas the Great Lakes' Environmental Era II does not begin until 1965 and ends in 1983. This is explained by wetlands being more severely impacted as a natural resource since most were completely eliminated early in the history of the U.S., leading to environmental concern beginning earlier than the Great Lakes. The point here is that the Great Lakes policy is following a dominance issue model profile like wetlands until Era III begins. The policy definition of ecosystem approach adapts to include a sustainable development outlook which gives room on the agenda for the navigation, industry, and commerce tone to reemerge from Era I. It is Era III in the Great

Lakes policymaking that makes a departure from the dominance issue model. The seeds of issue definition are set in Era II, but adaptation to allow agenda sharing of tones that occur help create a bounded issue model that develops in Era III.

Era III—Agenda Sharing (1984–1999)

Era III begins to fully implement many of the environmental policies and human health programs for the Great Lakes created in the mid-1970s. The programs created in the late 1970s were now, in the 1980s, institutionalized and having an impact on the basin. Botts and Muldoon (1996) have studied the effectiveness of the Great Lakes water quality regime created in Era II. They conclude that the Great Lakes are greatly improved from previous eras. The cooperation and agenda sharing, made possible by the issue definition of the Great Lakes changing, made this possible. Today anglers seek fishing spots at Lake Erie, the lake proclaimed dead for aquatic life. Some of this is attributed directly to the cooperation by users of the Great Lakes guided by the Great Lakes Water Quality Agreement in 1972 and later amendments.

For the first time in the Great Lakes Congressional policymaking, no tone dominates. A total of 127 hearings are conducted by 17 committees and 31 subcommittees. There are 557 witnesses in Era III who testify before Congress. Clearly environmental committees (32%) still maintain a strong hold on Congressional policymaking; however, there are a number of navigation, industry, and commerce committees conducting hearings (collectively 25%). Neither tone for the Great Lakes dominates the hearings being conducted as they once did in Eras I and II. The reliance on legislative hearings is still strong (76%), but is declining from the eighty-six percent reliance on legislative hearings from Era II. The federal government agencies (25%), especially the EPA, tend to remain frequent witnesses in Era II and Era III. However, hearings are more reflective of a sharing or perhaps cooperating relationship between the two tones. The Merchant Marines and Fisheries Committee dominates with thirty percent of all the hearings and all other committees are less than thirteen percent.

This era reflects a different path from the one that could have been taken that would mirror the wetlands case. After the dominance of Era I by industry, commerce, and navigation, the redefined tone for environment and health virtually became the dominant tone. This shift is not surprising based on the symbols like a dead Lake Erie or a burning river in Ohio. What is unique is the path taken from Era II to Era III. Unlike wetlands, the environmental and health committees and participants do not come into severe policy conflict with the resurrection of industry, commerce and navigation interests that re-emerge in Era III. Unlike wetlands, Era III results in a shared agenda, where the different definitions and tones of the Great Lakes policy coexist. There is a concern for water quality, quantities, and economic development in the management of the Lakes. There is no major debate or conflict, but some mutual ecosystem benefits from approaching the management of the Lakes from a sustainability function.

The Great Lakes Water Quality Agreement originally signed in 1972 is again amended in 1987 to include the identification of local areas of concern around the Lakes that had been significantly degraded and in need of remediation. Forty-three areas are originally identified for remediation. Remedial actions plans are developed and implemented at these sites. Twenty-six of the areas of concern are entirely within the U.S. and twelve located wholly in Canada. The remaining five sites are shared by both countries. Focus is placed on nonpoint sources of pollution, contaminated groundwater and sediment, and airborne toxins that were transferring into the Lakes. The International Joint Commission, U.S. EPA, Canadian provinces, Indian nations, and municipal governments along with Environment Canada must work together as partners to implement the agreement. The international community regards this partnership as a successful model for interjurisdictional cooperation toward restoring environmental quality and preventing future degradation through an ecosystem management approach (The Great Lakes Water Quality Agreement Backgrounder 2000).

By 1990, the population of the basin reaches 33.4 million, only eight percent more than in 1970. In the U.S., there are reports of behavior difference found in New York infants whose mothers ate Great Lake fish

(The Great Lakes Water Quality Agreement Backgrounder 2000). This keeps the issue of the Great Lakes environmental and health tone on the agenda. Colborn et al. (1990) provide a comprehensive overview of the implications for human health from the presence of toxic contaminants in the Great Lakes. In 1995, the U.S. EPA issues water quality guidance for the Lakes called the Great Lakes Water Quality Initiative. This includes the passage of standards by EPA in addition to regulations passed by states to meet those standards.

There are still environmental and health problems with the Lakes today. For instance, Lake Superior is suffering from lead, mercury, PCBs and DDT. Most of this pollution is not discharged into the waters but into the air. These persistent airborne substances migrate to Lake Superior and concentrate in the aquatic life. But, communities and the EPA are working together to cleanup and maintain the Lakes through remedial action plans for cleanup of the areas of concern and through lake-wide management plans. The pulp and paper industries of Lake Superior have improved, as well as the processing of wastewater treatment methods.

Botts and Muldoon (1996) characterize the Great Lakes policy as coordinated regional action that recognizes the economic and environmental importance of the Great Lakes, that transcends political boundaries, and focuses on an integrated ecosystem to achieve protection of the Lakes. Today, the Great Lakes agenda is clearly today one of economics and environment due to the inevitable linkages between the two tones within the basin (Allardice and Thorp 1995).

Testing the Relationships among the Great Lakes Policymaking Venues, Witnesses, and Hearings from 1789 to 1999

Congressional policy dynamics have a significant role in helping promote a certain type of the Great Lakes policy. Congressional committees engage in this process through three main routes. First, committees influence the definition or redefinition of a policy by conducting hearings with a certain tone. That is, instead of considering issues of the opposing tone, committees only conduct hearings that reflect their tone. The topic being considered usually is not neutral, and the subject of the hearing not open to consider a wide range of alternatives. Second, witnesses who testify before a committee can either reflect the tone of the committee or an opposing tone. Committees concerned about maintaining or expanding their jurisdiction have an incentive to select witnesses that have a corresponding tone. These two dynamics actually reinforce each other by satisfying the needs of policy advocates who need to venue-shop because their interests are not being considered by the status quo in Congress and by committees seeking to expand their jurisdiction. To investigate if these Congressional dynamics exist in the Great Lakes policymaking, a correlation should occur between tone of the hearings and committees, and between tone of the committees and the witnesses. To investigate venue-shopping and Congressional expansion, two tests were conducted using Chi-Square and correlation coefficients of Gamma and Kendall's Tau-b. In addition to the two categories of tone, a third category of other or was used for hearings and committees that were neutral in tone or uncodeable.

The first test was a crosstabulation between the tone of the committee's jurisdiction and the tone of the hearing. Table 4.2 shows that the committees tend to conduct hearings that reflect their definition of the policy (Chi-Square = 525; $p < .00001$; Gamma = .96 and Kendall's Tau-b = .86, $p < .00001$). Environmental committees tend

Table 4.2 Relationship between Tone and Venue in Congressional Hearings on Great Lakes

Venue	Tone of Hearings			
(Committees)	Industry, Commerce, and Navigation Tone	Environment and Health Tone	Neutral or Uncodeable Tone	Totals (N)
Industry, Commerce, and Navigation Tone	86% (117)	10.3% (14)	3.7% (5)	100% (136)
Environment and Health Tone	1% (1)	95.1% (98)	3.9% (4)	100% (103)
Neutral or Uncodeable Tone	1.4% (2)	12.9% (18)	85.7% (120)	100% (140)
Totals	31.7% (120)	34.3% (130)	34% (129)	100% (379)

Gamma = 0.96, Kendall's Tau-b = 0.86, Chi-Square = 525 ($p < 0.00001$)

to conduct hearings having an environmental tone versus other interests (95.1%). Likewise, industry, commerce, and navigation committees tend to conduct hearings with a tone reflective of their interests
(86%). Similar to wetlands, the environmental and health venues
tended to not be accessible to venue-shoppers with the industry,
commerce, and navigation tone. Only one percent of the hearings by
environmental and health Committees heard from the opposing tone.
The industry, commerce, and navigation venues were also generally
not available to the environmental and health witnesses. Industry,
commerce, and navigation committees conducted 10.3 percent of the
hearings with an opposing tone. Thus, this data shows that tone of
hearings and committee jurisdictions are excellent ways to understand
how Congress helps defines the Great Lakes policy. Committees are
using hearings to garner support for the views they already hold and
establish their jurisdiction over a certain aspect of the Great Lakes
policymaking.

    Like committees who hold hearings to help expand and defend
jurisdictional claims to the Great Lakes policy, committees can select
witnesses to reinforce the committee's jurisdictional tone. From the
analysis of the three eras of the Great Lakes policymaking, no single
witness or representative group had a large monopoly over another testifying before Congress, which is similar to the wetlands case. However,
Table 4.3 shows that committees conducting hearings on the Great
Lakes policies hear from interests much like their own interests
(Chi Square = 1833; $P < .00001$; Gamma = .81 and Kendall's Tau .60,
$p < .00001$). Industry, commerce, and navigation committees have
slightly more of a tendency (86.3%) to hear from witnesses much like
themselves than environmental committees (83.1%). Industry,
commerce, and navigation committees only heard from opposing
witnesses 9.6 percent and environmental and health committees 15.1
percent. Thus, Congressional committees can use hearings and
witnesses to help express their particular tone on the Great Lakes. Also,
venue-shopping by the Great Lakes witnesses appears to be an opportunity for a policy loser to seek out committees sharing their tone.

    The third Congressional dynamic that assists with policy changes is

**Table 4.3 Relationship between Venues and Witnesses
in Congressional Hearings on Great Lakes**

Venue	Witness Tone			
(Committees)	Industry, Commerce, and Navigation Tone	Environment and Health Tone	Neutral or Uncodeable Tone	Totals (N)
Industry, Commerce, and Navigation Tone	86.3% (2071)	9.6% (230)	4.2% (100)	100% (2401)
Environment and Health Tone	15.1% (118)	81.7% (639)	3.2% (25)	100% (782)
Neutral or Uncodeable Tone	22.7% (71)	67.4% (211)	9.9% (31)	100% (313)
Totals	64.6% (2260)	30.9% (1080)	4.5% (156)	100% (3496)

Gamma = 0.81, Kendall's Tau-b = 0.60, Chi-Square = 1722 ($p < 0.00001$)

the relationship between Congressional attention and venues. As Congressional attention on the Great Lakes increases, the number of venues claiming jurisdiction also should increase. Congressional attention to the Great Lakes attracts new venues to capture jurisdictions that are unclaimed by others. An ordinary least squares regression, used to test how much Congressional attention is paid to the Great Lakes, explains the growth of venues claiming the Great Lakes jurisdiction. Equation 4.1 shows the ordinary least squares equation used in this analysis.

Equation 4.1 Congressional Attention and the Role of Committees

$$Y_t = \beta_1 X_t + \beta_2 Y_{t-1} + e_t$$

Y_t = Number of Hearings conducted each year on the Great Lakes. (dependent variable)

X_t = Congressional Committees holding Hearings on the Great Lakes each year.

Y_{t-1} = Number of Hearings conducted previous year on the Great Lakes (lag variable).

e_t = residual unexplained by model.

Congressional attention, measured as the number of hearings on the Great Lakes per year (independent variable), results in increases to the number of available Great Lakes venues as measured by the number of subcommittees (dependent variable). An endogenous lag variable was included to control for hearings conducted the previous year on the Great Lakes. This lag variable allows the regression model to reflect actual growth in Congressional attention that cannot be explained by the previous year's hearing activity.

The regression results in Table 4.4 indicate that as Congressional attention increases, Congressional venues proliferate to accommodate venue-shoppers and committees seeking to expand their jurisdiction (adjusted R Square = .77; Beta = .82; p < .0001). In fact, the results show that Congressional attention has a large impact on the expansion of venues. A nice triangulation of results is demonstrated when Table 4.1 comparing the eras with the Chi-Square tests and regression analysis and all are taken into consideration. All results point to committees and witnesses using Congressional hearings for redefining policymaking on the Great Lakes.

The Great Lakes as Bounded Issue Model

Congressional Great Lakes policymaking has gone through significant changes. These changes can be classified into three eras based on a dominance of particular interests reflected in Congressional hearings.

Table 4.4 Relationship between Congressional Venues and Hearings in Wildlife from 1789 to 1999

Variable	B	Standard Error	Beta Coefficient	Significance
Intercept	0.49	0.134		< 0.0001
Wildlife Hearings	0.28	0.03	0.69	< 0.0001
F Statistic				< 0.0001
Adjusted R-Square	= 0.761			
*Durbin Watson**	= 2.20			
N	= 112			

*Indicates no autocorrelation problem. The dependent variable is annual number of subcommittees holding hearings on wildlife.

Until recently, the Great Lakes were seen as a natural resource to be used for supporting industry, commerce, and navigational interests of the United States. Evidence for this conclusion comes from the examination of committees and witnesses who testify before Congress. In Era I (1789–1965), Congressional policymaking defined the Great Lakes as a resource needed to support the economy of the country and the surrounding Great Lakes region. The dominance of navigation, industry, and commerce is strong with nearly one hundred percent of the hearings being dedicated to this tone until 1965.

In Era II, which begins in 1966 and ends in 1983, the previous domination by navigation, industry, and commerce is replaced by the concerns for environment and health brought onto the agenda by the environmental movement in the United States. Policymaking focused more on defining the Great Lakes as a natural resource that needed protection from pollution and development. Instead of hearings focusing on building ports, canals, or providing for shipments of commodities as in Era I, the hearings reflect concerns about pollution levels and misuse of the Great Lakes, an important environmental resource. This new dominant tone of the Great Lakes does not last as long as did the navigation, industry, and commerce dominance of Era I. Since 1984, Congressional attention has not been dominated one hundred percent by either tone. Certainly, concern over the environmental and health aspects regarding the Great Lakes has not been substituted. Nonetheless, Congressional attention has been focused on both the tones from Era I and Era II. Today, the Great Lakes are defined in Congress as having both industrial and environmental tones.

Era III (1984–1995) can be labeled as agenda-sharing. On the one hand, environmental and health interests lost the monopoly of Congressional attention and now must share that attention with industry. On the other hand, this could reflect a more balanced or ecosystem approach to the Great Lakes policymaking in Congress. In other words, today Congress views the Great Lakes as a resource for economic gain as well as one that needs environmental and health protection.

Unlike wetlands policy that literally eliminated the natural resource, the impacts of Era I in the Great Lakes, to a certain extent, are being reversed. This may also explain the more intensely conflictual route wetlands policy took as a dominance issue model. Today, the Great Lakes agenda includes a wide variety of interests that in Era I and II were not seen as compatible. This linkage between industry and environmental protection allows a bounded issue model to exist through a mutually beneficial relationship captured in Era III's ecosystem and sustainable development issue definition of the Great Lakes. Water quality of the Lakes is improving while the basin is experiencing economic growth in the 1990s. Many layers of government and participants have come together to address emerging air pollution issues around the basin which impact the Lakes. Even recreation has grown steadily in recent years (U.S. EPA 1995). Current challenges to this bounded issue are water diversion. At one point, flooding and erosion were major concerns of property owners along the basin. As of the last three years, the weather patterns have caused a six-inch decrease in water levels. Instead of problems historically associated with high lake levels, the new challenge is to develop policies to deal with unusually low levels of lake water. The joint management of the Great Lakes continues to reflect the benefits of a bounded issue model, which has shown it actually restores a natural resource. The next chapter looks at a different path an issue can take and how it can impact the natural resource.

CHAPTER 5

Valence Issue Model—
The Case of Wildlife

Wildlife policy has had a more recent history of policymaking than other environmental issues like the Great Lakes that can date back to the 1600s when the Lakes were discovered. Wildlife policy is no less important, but generally did not capture the attention of Congress until late the1800s. Endangered species is one aspect of wildlife policy that has been recently considered from an economic point as a public good (Fredmand and Boman 1996). Therefore, like wetlands and the Great Lakes, wildlife policy has the similar divisions of private development versus environmental protection at the core of its issue definition. However, as the details of wildlife policy are explained this chapter, it becomes clear that this issue is completely different from wetlands or the Great Lakes. Wildlife as a policy reflects a valence issue model.

Relationship in Wildlife Policymaking of Congressional Hearing and Committee Dominance

Using the CIS on CD-ROM for Congressional hearings from 1789 to 1999, a search was performed using the keyword wildlife. To avoid missing hearings on wildlife, synonyms such as threatened species, endangered species, and extinction were tracked along with conservation. This search yielded a total of 466 hearings with 4500 witnesses and reflects the complete set of Congressional hearings on wildlife policy in the United States. Even with this comprehensive tracking, no hearings were detected prior to 1888.

Tone of committees, witnesses, and hearings were recorded as a dichotomous variable. For wildlife policy, tone was defined as either representing private development or trade issues and environmental protection issues. For coding environmental protection tone, issues included concerns about protecting and preserving wildlife (including their habitat for both plant and animals) as a natural resource. Often the environmental protection tone included environmental interest groups and sometimes recreational groups interested in hunting, sports, and leisure. For coding private development and trade tone, issues included being favorable toward agriculture, private property owners and developers, and parties interested in trade or commerce involving species. Since the divisions between these groups in the wildlife policy are distinct, it was a valid classification to make in the coding effort. Hearings and witnesses were coded separately from the committee based on the information provided by CIS. This technique allowed comparisons between tone of the venue with tone of the hearing and witnesses. Key issues of concern by witnesses were also tracked over time.

Congress made wildlife policy using thirty-three committees and seventy-three subcommittees. Figure 5.1 shows the number of Congressional committees that held annual hearings on wildlife policy.

Generally, the trend in this figure shows that wildlife hearings have increased gradually over time since the early 1900s. There is a more dramatic increase in the number of hearings in the early to mid-1990s with a recent decline in the late 1990s. Figure 5.2 shows the number of witnesses that testified annually on wildlife policy. There is an increase in witnesses participating in hearings in the 1960s, and noted increase in the early 1990s with a decline in the mid-1990s. Both of these data sets are used for the analysis of wildlife policy. Like Figure 5.1, there is a rather continuous participation in the number of hearings and witnesses over time. In general, Figures 5.1 and 5.2 show a continuous number of hearings and witnesses since mid-1900s with somewhat of an increase occurring in the 1990s. Besides tone, this information was coded for supplemental data information, such as gender of witnesses and organizations represented. Individual codes

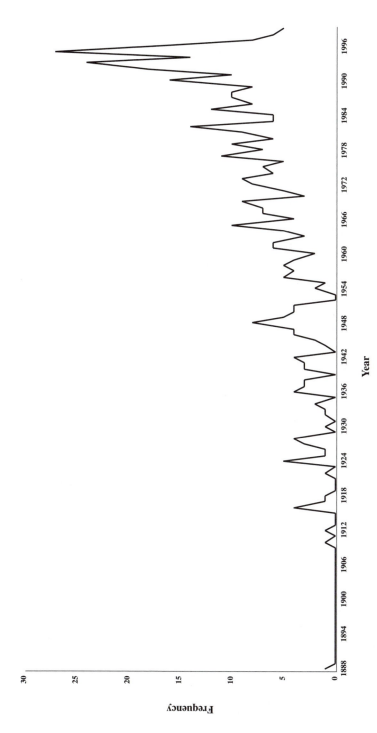

Figure 5.1 Congressional Hearings on Wildlife

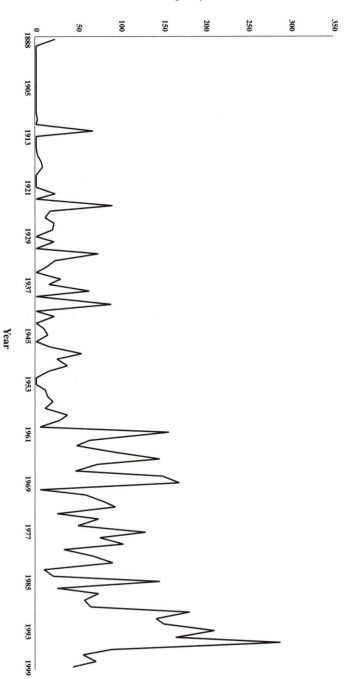

Figure 5.2 Witnesses Testifying on Wildlife

for each witness were used to track if witnesses testified multiple times indicating a reliance on certain interests by Congress.

In addition, hearings were coded as legislative or nonlegislative. Hearings that consider bill-referrals were defined as legislative. Nonlegislative hearings are particularly important because few restrictions exist on the subject of hearings. Generally, committees tend to use nonlegislative hearings to encroach on the jurisdictional control of another committee that would challenge the committee's monopoly. The next section on results describes in detail how wildlife issues have enjoyed a continuous positive definition in Congress over time.

Results

Wildlife Congressional Policymaking

Of the 466 total hearings conducted, most considered formal legislation (71 percent). For the total set of hearings, the majority had a tone that was focused on environmental protection of the species (57.7 percent). Only eighteen percent had a tone that contained support for private development and trade. Many of the hearings were conducted as appropriations so it is not unusual to have approximately twenty-four percent be in the neutral category for tone. Unlike hearings on either the Great Lakes or wetlands, there is not any dominant tone for hearings or witnesses that would indicate separate eras of policymaking on wildlife.

By examining the annual percentage of hearings in the environmental protection venue versus the private development venue, no trends emerge over time. Unlike wetlands and the Great Lakes, the tone of the issues has not been significantly changed over time. Table 5.1 shows the data for the policymaking on wildlife policy over time. This table illustrates the venue and witness information presented in the following results section.

Wildlife Policy (1789–1999)

Development of federal wildlife policy is not an issue that began in the 1960s and 1970s environmental era. Ernst (1991) traces the beginnings

Table 5.1 Congressional Wildlife Policymaking from 1789–1999

A. Hearing Information

Number of Hearings	466
Number of Committees	33
Number of Subcommittees	73

Percentage of hearings held by each Committee

Merchant Marines and Fisheries	23%
Environment And Public Works	7.3%

All other Committees accounted for 6%
or less of the hearings conducted.

Percentage of hearings held by each Subcommittee

Fisheries & Wildlife Conservation	15%

All other Subcommittees accounted
 for 4% or less of the hearings conducted.

Type of Hearing

Legislative	71%
Nonlegislative	29%

Percentage of Type of Committees

Environmental	50%

All other types of committees accounted
 for less than 9% of the committees.

B. Witness Information

Number of Witnesses	4500

Percentage Gender of Witnesses

Male	84%
Female	10%
Uncodeable	6%

Percentage of Witnesses that testify before each Committee Environmental — 61%

All other witnesses accounted
for less than 8% of the committees.

Percentage of witnesses testifying before each Subcommittee

Fisheries and Wildlife
Conservation 22%
All other witnesses accounted
for less than 6%.

Percentage of Selected Witnesses Specific Witnesses Testifying

Federal Government	24%
Interest Groups	20%
Trade Associations	12%
Federal Elected Officials	9%

All other witnesses accounted for less
Than 6% of those that testified.

of U.S. wildlife policy back to an early definition from the Roman Empire, feudal Europe, and the Magna Carta. At the time of the American Revolution, England's king exercised complete control over wildlife, which was considered a public trust. As early as 1842, the U.S. Supreme Court entrusted the state government with authority to regulate wildlife once held by the king (*Martin v. Wadell*; 41 U.S. (16 Pet) 367 C 1842). This court ruling served as the basis for state ownership which reflected an environmental tone that guided wildlife policy throughout the nineteenth century. Declining populations of wildlife in the 1800s prompted states to create wildlife agencies. Efforts to maintain the heath hen prior to the Civil War, protect bison in the 1870s, and establish the Aransas National Wildlife Refuge in 1973 to protect whooping cranes are examples of early wildlife issues. These early efforts defined wildlife with an environmental tone. At the end of the nineteenth century, states' role as wildlife managers was supported by challenges in the courts. However, Ernst documents that with the birth of conservation management in the early twentieth century significant challenges to state control were made by Teddy Roosevelt, John Muir, John Audubon, and the Congress by holding hearings which passed federal wildlife legislation, beginning with the Lacey Act of 1900. Several major pieces of federal policies were enacted in the 1900s. Kohm (1991:10–31) describes the early definition of wildlife in the United States. Until 1900, jurisdiction over wildlife remained largely with the states where conservation efforts were focused almost exclusively on traditional game species. By the late 1800s, well-organized commercial interests had overrun the efforts of individual states to enforce their wildlife laws by killing enormous quantities of wildlife in one state and quickly transporting them to another. As early as 1888, Congress held hearings with testimony of twenty-three witnesses of concern regarding wildlife protection. By 1895, hunting was prohibited in Yellowstone National Park. Near the turn of the century, the balance of power over wildlife began to shift from the state and local governments to the Federal government. In response to the rapid decline of passenger pigeons, Congress passed the Lacey Act of 1900, making a significant, although cautious, entry into the field of wildlife regulation (Bean 1983;17). Using the commerce clause of the Constitution, Congress passed the Lacey Act to bolster enforcement of existing state wildlife regulations. In the series of federal wildlife laws that followed the Lacey Act, elements of the current federal endangered species program began to emerge. For example, in 1906, hunting of birds on federal lands reserved as breeding grounds was prohibited.

What is unique about the definition of the wildlife issue is that it continuously remains one of protection with some challenges from private development and trade which do not result in significant redefinitions of tone. Conflict focused on what level of government, that is federal or state, would be the defender and manager of wildlife: This organizational conflict has not impacted the basic premise which defines wildlife today, that it is valued and should be protected. Soon after the Lacey Act, there are a series of hearings to establish protection of game birds by Congress. This effort produced the 1918 Migratory Bird Treaty Act. However, from 1789 to 1999, there were four pieces of major legislation dealing with wildlife issues which all reflect a monotonic of protection. In addition to the Lacey Act of 1900, an even stronger environmental protection tone of wildlife policies follow in the Endangered Species Act of 1966, the Endangered Species Conservation Act of 1969, and the Endangered Species Protection Act of 1973.

The Lacey Act prohibited the transportation of interstate commerce of animals killed in violation of state laws. This was expanded in 1935 to include any animals taken in violation of foreign laws as well. This helped to protect species because state and foreign laws were supplemented by federal penalties and enforcement. Also, this early wildlife protection policy included that a permit be required before any wild animal or bird be imported into the United States. Animals and birds that were shipped had to be transported under humane conditions. Originally the Lacey Act helped wildlife protection, but was intended to prevent importation of animals injurious to agriculture. The Lacey Act was the first federal effort to regulate wildlife that had been considered a state responsibility. Part of the reason this Act was initiated was the decimation of the passenger pigeon, a clear symbol of the need for protection of species. According to Palmer (1975; 258), the Lacey Act was recognized in 1900 as designed to protect wildlife. A major deficiency that led to its ineffectiveness is that it was dependent upon local and foreign laws for its usefulness, rather than embodying a substantive federal program designed to ensure the conservation of species (Coggins 1973).

There were many other legislative programs, such as the ones created under the Migratory Bird Treaty Act of 1918, the Migratory Bird Conservation Act of 1929, the Fish and Wildlife Conservation Act of 1934, the Bald Eagle Protection Act of 1940, the Free Wild Roaming Horses and Burros Act of 1971, and the Marine Mammal Protection Act of 1972. All of these pieces of legislation, along with numerous preserves and wildlife reservations created by Congress, clearly defined wildlife as an issue with a very protective tone. The established tone from the beginning for wildlife was a public trust to be preserved and protected. This is completely different than the original tones established in Era I for wetlands and the Great Lakes policies. And, more interesting, this strong tone for wildlife maintains itself as the dominant tone since the first Congressional hearing in 1888.

For instance, habitat protection began as early as 1903 with the designation of Pelican Island National Wildlife Refuge. From 1900 to 1920, Congress established refuges throughout the country. By 1929, Congress established a commission under the Migratory Bird Conservation Act to review the Department of Interior's proposals for refuge purchases. The 1934 Migratory Bird Hunting Stamp Act and Pittman-Robertson Act assist with the funding for refuge acquisitions. The Fish and Wildlife Coordination Act of 1934 set a precedent for requiring federal agencies to consider the effect of their actions on wildlife populations and advocated intergovernmental cooperation to develop a national conservation program.

According to Kohm, the maturation of federal wildlife law was propelled by increasing concern among wildlife professionals and the general public over loss of species. By the environmental movement of the 1960s, wildlife was defined as environmental protection. The Committee on Rare and Endangered Species in the Interior Department's

Bureau of Sport Fisheries and Wildlife was established in 1964. This committee was composed of nine biologists who published the first endangered species list called the "Redbook." This list created more national media attention and awareness, which added to the passage of the future wildlife protection laws in 1966, 1969 and 1973.

The Endangered Species Act of 1966, was the first domestic law exclusively concerned with the protection of endangered species. It focused on habitat destruction and native wildlife. Under this legislation, a National Wildlife Refuge System was established under the Department of Interior for the conservation of fish and wildlife. This Act confined its prohibitions to action within an established refuge and allowed hunting, capture, and exploitation of endangered species outside the refuge where applicable local laws were absent. The focus was on habitat destruction and native wildlife, which ignored the international aspect of endangered species problems.

Within a couple years, Congress passed another significant piece of legislation to protect the endangered species from over commercialization. In 1969, the Endangered Species Conservation Act amended some of the deficiencies of the 1966 programs. By this time, Congress acknowledged that commercialization was a major cause of extinction. Therefore, import controls, permits, and other tools were used to prohibit the market for endangered species and their manufactured by-products. The Secretary of the Department of Interior was required to convene an international meeting to assist in regulating worldwide trade. This resulted in the 1973 Convention on International Trade in Endangered Wildlife, Fauna, and Flora Agreement. The response to this treaty was the passage of the Endangered Species Act (ESA) of 1973, which was patterned after the treaty. This 1973 Act superseded the 1969 legislation, replacing all of it except for the National Wildlife System.

Under the 1973 legislation, the issue of species protection had expanded to include protection of not just endangered species but also threatened ones. This was significant expansion of the issue of species protection because now the Department of Interior had the power to protect animals before they became endangered. This species protection legislation is considered a pro-environmental stance, which provides for increased public participation, increased scope of civil and criminal liabilities, and penalties which clearly indicate a lack of neutrality in tone (Coggins 1973). Several exemptions were given based on determinations of commercial versus noncommercial uses and the legal concept of "taking." According to Dingell (1991; 25) who served as chairman of the Subcommittee on Fisheries and Wildlife Conservation and Environment of the House Committee on Merchant Marine and Fisheries that introduced the bill, its goal was unparalleled in all history. Extinction of the passenger pigeon and Carolina parakeet, as well as near extinction of the bison from decades earlier, were all symbols that conveyed support for the expansive ESA of 1973. Other even more powerful symbols included the bald eagle that was close to extinction because of the pesticide DDT.

Of course, wildlife policy is not without conflict. Valence issues may be challenged with tests to the definition of the issues, particularly when property issues are involved. The key to a valence issue is that it endures the challenges due to its definition neutralizing the mobilization of opposition. A valence issue does not incorporate or adapt its tone as does a bounded issue model. A valence issue does not use symbols or rhetoric that polarize oppositional tones. Instead, it uses powerful neutralizing symbols and rhetoric that prevent an opposing tone from being effective in the redefinition process. For instance, Dingell gives a first-hand account of the endurance of ESA due to the ethics and symbolism involved with protecting wildlife. One example of this endurance occurred in 1978 with one of the first challenges to ESA: the construction of the Tellico Dam under the Tennessee Valley Authority. Near completion, the construction of this dam was halted due to a rare fish called the snail darter. The discovery of the snail darter, which was listed as an endangered species, helped property owners concerned about the impact of the project to property values delay the project. In January 1977, the Sixth Circuit Court of Appeals upheld the ESA authority to stop construction. Tellico developers appealed the case to the Supreme Court, who upheld the lower court decision in June 1978. Congressional elected officials from Tennessee along with developers tried to eliminate portions of the ESA that prevented destruction of habitat of an endangered species. A compromise was reached to create an independent board to resolve conflicts that revolved around species habitat. The first decision of this commission was against the Tellico Dam project. However, in 1979, Congress passed a measure to continue construction of the dam by attaching a rider to a House energy and water appropriations bill. The Senate narrowly voted to continue the project after the snail darters were removed to nearby waters.

Another major challenge to the ESA came from western water rights issues. Tarlock (1991) points out that amid these powerful coalitions in the western states, ESA has not weakened and is unlikely to do so in the future in light of increasing appreciation for biological diversity. The author points out that ESA is a significant environmental constraint that affects western water rights. The issue is not whether water rights for endangered species exist, but under what circumstances and in what manner they can be asserted (Tarlock 1991; 173). The author gives several case studies on how the FWS has successfully used ESA to create federal water rights.

The gray wolf has also challenged the ESA. Beginning in the late 1800s, the animal that conservationist and president Theodore Roosevelt called the "beast of waste and desolation," was subjected to a government sponsored extermination program. Between 1870 and 1877, government sanctioned hunters killed 55,000 wolves every year. In 1914, the federal government hired hundreds of hunters to kill predators, including all wolves. By 1926, rangers had killed at least 136 wolves in Yellowstone Park alone (Zuccotti 1995; 330). Today the wolf survives in one percent of its original range in the lower forty-eight states. The wolf was listed in 1973 as an endangered species as an effort to reintroduce the wolf into Yellowstone National Park. This reintroduction has caused some controversy with local farmers and property owners who are negatively impacted from the reintroduction of this negatively portrayed predator. Unlike other charismatic animals that have benefited and helped the wildlife policies in the United States, the wolf has not enjoyed such a positive image. In 1987, the FWS undertook a Gray Wolf Recovery Plan in the Northern Rockies that also had controversy. In May 1994, the FWS released its study for the reintroduction of experimental wolf populations in Yellowstone Park and Central Idaho. Wolf recovery is expected for these areas by year 2002. Reaction to the FWS plan has been accepted but with strong concerns expressed by ranchers about the potential loss of livestock.

ESA has been amended in 1978, 1982, and 1988 with the overall framework of the 1973 Act essentially remaining unchanged. Each amendment added to the scope of species protection as a policy issue. For instance, in 1978 critical habitat was required to be defined concurrently with the listing of a species. Both the FWS and Forest Service (FS) were directed to develop programs for conservation of fish, wildlife, and plants where land acquisition authority was extended to such species. In 1982, amendments included that determination of the status of species were required to be made solely on the basis of biological trade information without any consideration of possible economic or other efforts. In 1988, recovery of species and emergency listing of species were defined and expanded.

By most accounts, ESA is one of the most powerful environmental laws of the century. One of the first jobs for the FWS after the ESA was signed was to gather statistics to show that since Tellico Dam there have been virtually no conflicts between endangered species and development (Greenwalt 1991). Unlike other environmental programs, this one is intimately connected to moral concern, issues of natural values, esthetics, and human protection of nonhuman species (Rolston 1991). These powerful symbols get extended into other international arenas with the Convention on International Trade in Endangered Species of Wild Fauna and Flora (CITES) where Congress constrains trade to protect plants and animals. Unlike wetlands, the definition of issues with wildlife does not have the long history of change. Indeed, property owners and developers must tangle with the issues of impacting or taking private property in both wetlands and wildlife issues. Even with the more recent spotted owl controversy in the Pacific Northwest, environmental protection prevailed in preventing the timber industry from destroying its habitat in the old growth forests (Yaffee 1994). However, wetlands has a long policy path of issue definition and redefinition which caused great harm to the natural resource when it was defined as swamps, bogs, and nuisance lands to farmers and developers. The Great Lakes policy was defined, redefined, then adapted to include multiple tones of the issue. Wildlife issues never had the overwhelming redefinition take place because they were defined from the origin as intimately tied to the nation's symbols for freedom, liberty, and prosperity. The FWS, under the Department of Interior, views hunting as a valuable tool for population control as part of its philosophy (Coggins 1991). The FWS along with the National Marine Fisheries Service has taken the primary lead on ESA with roles for policy also including the Forest Service, Bureau of Land Management, National Park Service, and EPA. Nonetheless, the FWS has a pro-wildlife tone to its implementation of wildlife that has persevered over time. National Marine Fisheries Service, under the Department of Commerce, deals with those species occurring in marine environmental and anadromous fish while the FWS is responsible for terrestrial and freshwater species, along with migratory birds.

Another major aspect to wildlife that assisted it as a valence issue is how conflict is managed. Wildlife programs include an interagency consultation provision, which is a concept of cooperation defined in wildlife law in the early 1970s. Section 7 of the ESA of 1973 strengthens the interagency consultation process, which provides a seldom used exemption process. Yaffee (1991; 87) describes how ingenious this consultation process is at dealing with neutralizing conflict. This consultative process, along with the commission used to grant exemptions, provides an essential political pressure valve that handles, deals with, and eliminates conflicts. This consultative process has been used to expand endangered species policy into protecting habitat rather than individual animals.

More evidence of the continued support for wildlife comes from the appropriations history of the FWS. Although, the FWS is one agency dealing with wildlife, its funding profile clearly supports its implementation of valence issues. When the appropriations are examined from 1971 to 1998, which include prelisting, listing, consultation, permits, recovery, and species conservation fund, there is a seventeen times real increase. Using 1998 constant dollars, funding for species protection at the FWS has increased from approximately $6.5 million dollars to (a little) over $112 million dollars (see Figure 5.3). There is a decrease that occurs in 1982 during the Reagan Administration to almost 1978 funding levels. However, that funding level rebounds to 1982 levels before the decline by 1986 during Reagan's second term in office. The FWS has not seen a significant decline since the Reagan Administration years from 1981 to 1986, but a steady growth with minor shifts in funding occurring in 1994 and 1997. The trend is clearly upward for support of wildlife by the FWS. Campbell (1991) documents the appropriation history for other agencies associated with wildlife that indicate similar, but less dramatic trends.

Today, efforts on wildlife, which have traditionally focused on the protection of individual species, have been altered to include a more

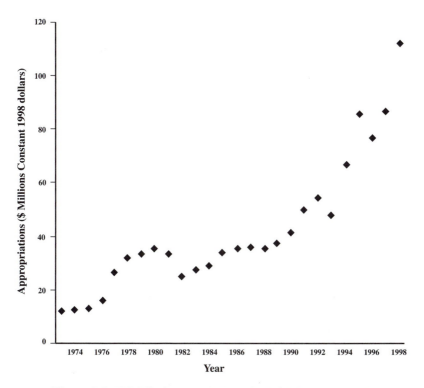

Figure 5.3 Wildlife Appropriations for ESA Implementation

expansive definition, which includes ecosystem management. This concept has been the approach for the Great Lakes policy. However, it has meant different things in each policy. In wildlife policies it allows a strengthening of the environmental protection tone. Sustainable development is not part of the issue definition for wildlife policy as it is in the Great Lakes. Charismatic species such as grizzly bears, spotted owls, bald eagles, and bison continue as strong symbols for wildlife protection. Ehrlich and Ehrlich (1981) illustrate the compassion intrinsic and aesthetic values place on wildlife. Since the 1880s, this use of visual symbols has supported a monotone of the wildlife issue definition. Description of American icons dramatically declining in population such as bald eagles, whales, passenger pigeons, gorillas, bison, and butterflies are visual symbols that have helped maintain this commitment by the U.S. to making wildlife a priority.

There have been conflicts and opposing viewpoints to wildlife policies, particularly those dealing more recently with endangered species protection. Simon and Wildavsky (1996) claim that extinction is a natural process that allows new species to evolve over time. While these authors point out the uncertainty in the scientific definition of a species and the role of extinction, clearly the issue definition of wildlife as embodied in policies and progress is one of environmental protection.

While the ESA is accused of unsound science based on altruistic values of nature (Knickerbocker 1995), experts frequently come to the aid of wildlife protection such as the National Academy of Sciences (Cone 1995). Wildlife protection receives criticism from scientists for not doing enough to recover species (Rohlf 1991) or being deficient in its implementation of its recovery planning (Cheever 1996). Criticism generally includes the overall design of the ESA that focuses too much on species rather than ecosystems, the impact on private property and the enforcement problems that stem from needing more funding (Kibasek and Browne 1994). Some criticisms claim the ESA does too much some say not enough. Some have even proposed revisions to the endangered species protection by transforming it into a tradeable program similar to the emissions trading on the stock market (Sohn and Cohen 1996). At the heart of most calls for revision to the

ESA is the notion of property rights for development and the private sector. While ESA has had some controversy, the core of its expanded protection has endured with little revision, except for strengthening wildlife protection over time (Sax 1997).

The most recent controversy, which involved presidential participation for resolution by Presidents Bush and Clinton, occurred in the Pacific Northwest old-growth forest, which is home to the spotted owl. This was a traditional property rights versus habitat protection issue for a single species. The issue was not adequately resolved under President Bush (Meyers 1991) and required intervention by President Clinton. The FWS began to use Conservation Agreements in the early 1980s then phased them out in 1985 with a reintroduction of them in 1992. This is a policy where the FWS uses agreements to substitute for listing a species on the threatened or endangered list. The agreement effectively removes all threats to the species that would otherwise warrant listing. This approach has been seen as a backlash from the spotted owl controversy to keep species off the list (Phelps 1997). Nonetheless, many accounts have been publicized and accepted as successes from ESA such as the resurging populations of bald eagles, whooping cranes, peregrine falcons, and elephants (Rohlf 1989). Even during the early debates in 1972 on ESA in Congress, there were no debates on the definition of wildlife protection, only on technical and administrative matters. Private property and trade interests infrequently testified and even in 1966, there were few opposed to ESA and wildlife protection. Rohlf points out that ESA was not viewed as opposed to any strong economic interests so controversy in the Congress was minimal. As a result, in July 1973, ESA was passed by unanimous vote in the Senate and by a huge margin in the House.

Like the Great Lakes policy, Table 5.1 shows that the committee holding the largest number of Congressional hearings from 1789 to 1999 is the Merchant Marines and Fisheries Committee (23 percent). Besides the Environment and Public Works Committee (7.3 percent), all other committees conduct less than six percent of all the hearings. The largest number of hearings conducted by a subcommittee is fifteen percent (Fisheries and Wildlife Conservation Subcommittee). All other subcommittees conduct four percent or less of the total hearings. While there is little controversy, clearly the largest Congressional agenda setter is the Merchant Marines and Fisheries Committee over time. The type of committee that dominates usually is focused on environmental tone (50 percent) with all other types of committees accounting for less than nine percent.

Based on the 4500 witnesses who testify on wildlife, Table 5.1 also shows that eighty-four percent are males. This is similar to the large majorities of males who testify in wetlands and the Great Lakes hearings. A majority (61 percent) of witnesses testify before environmental committees with all other subcommittees accounting for less than six percent. Not surprisingly, it is the Federal government (24 percent) and interest groups (20 percent) that are the largest participants appearing before Congress. Trade associations (12 percent) and Federal elected officials (9 percent) accounted for a smaller percentage with all other types of witnesses accounting for less than six percent.

Testing the Relationship among Wildlife Policymaking, Witnesses and Hearings from 1789–1999

In both wetlands and the Great Lakes policies, Congressional policy has had a significant role in helping promote a certain type of issue definition. In wildlife policy, there has been a continued influence of environmental protection as the definition of the issue. Nonetheless, Congressional committees engage in this issue definition process through three main routes, which are tested for wildlife issues. First, committees influence the definition or redefinition of a policy by conducting hearings with a certain tone. That is, instead of considering issues of the opposing tone, committees only conduct hearings that reflect their tone. The topic being considered usually is not neutral and the subject of the hearing not open to consider a wide range of alternatives. Second, witnesses who testify before a committee can either reflect the tone of the committee or an opposing tone. Committees concerned about maintaining or expanding their jurisdiction have an incentive to select witnesses that have a corresponding tone. These two dynamics actually reinforce each other by satisfying the needs of policy advocates who need to venue-shop because their interests are not being considered by the status quo in Congress and by committees seeking to expand their jurisdiction. To investigate if these Congressional dynamics exist in wildlife policymaking, a correlation should occur between tone of the hearings and committees, and between tone of the committees and the witnesses. To investigate venue-shopping and Congressional expansion, two tests were conducted using Chi-Square and correlation coefficients of Gamma and Kendall's Tau-b. In addition to the two categories of tone, a third category of other was used for hearings and committees that were neutral in tone or uncodeable.

The first test was a crosstabulation between the tone of the committee's jurisdiction and the tone of the hearing. Table 5.2 shows that the committees tend to conduct hearings that reflect their definition of the policy (Chi-Square = 598; p < .00001; Gamma = .96 and Kendall's Tau-b = .82, p < .00001). Committees with an environmental protection tone tend to conduct hearings having an environmental tone versus other interests (94.5 percent). Likewise, private development and trade committees also conduct hearings with a tone reflective of their interests (68.2 percent). The environmental protection hearings were sometimes also conducted by the private development and trade committees (29.9 percent). This reflects the valence nature of wildlife issues. Approximately one-third of the time the opposing tone to environmental protection entertained hearings not unlike their own. Thus, this data shows that tone of hearings and committee jurisdictions are excellent ways to understand how Congress helps define wildlife policy. Committees are using hearings to garner support for the views they already hold and establish their jurisdiction over a certain aspect of wildlife policymaking. Unlike the other cases of wetlands and the Great Lakes policymaking, wildlife issues exhibit a strong tendency toward environmental protection as a tone over time.

While these results are similar to the findings of Jones et al., what is different in wildlife hearings is who testifies. From the analysis of

**Table 5.2 Relationship between Tone and Venue
in Congressional Hearings on Wildlife**

Venue	Tone of Hearings			
(Committees)	Private Development and Trade Tone	Environmental Protection Tone	Neutral or Uncodeable Tone	Totals (N)
Private Development and Trade Tone	68.2% (73)	29.9% (32)	1.9% (2)	100% (107)
Environmental Protection Tone	4.2% (10)	94.5% (225)	1.3% (3)	100% (238)
Neutral or Uncodeable Tone	1% (1)	9.9% (12)	89.3% (108)	100% (121)
Totals	18.0% (84)	57.7% (269)	24.2% (113)	100% (466)

Gamma = 0.96, Kendall's Tau-b = 0.82, Chi-Square = 598 (p < 0.00001)

wildlife witnesses, no single witness or representative group had a large monopoly over another testifying before Congress. Likewise, Table 5.3 shows that committees conducting hearings on wildlife policies hear from interests unlike their own (Chi-Square 227; p < .00001; Gamma = .29 and Kendall's Tau-b = .15, p < .00001). Environmental committees have more of a slight tendency to hear from environmental witnesses (63.8 percent) than private development and trade related committees (42.5 percent). Private development and trade committees were more likely to consider an opposite viewpoint from the environmental protection tone witness (45.5 percent) than were environmental committees willing to hear from an opposing tone (25.7 percent). Clearly the overall number of witnesses (60 percent) and hearings (57.7 percent) were from an environmental protection tone.

The third Congressional dynamic that assists with policy changes is the relationship between Congressional attention and venues. As Congressional attention on wildlife increases, the number of venues claiming jurisdiction also should increase. Congressional attention to wildlife attracts new venues to capture jurisdictions that are unclaimed

**Table 5.3 Relationship between Venues and Witnesses
in Congressional Hearings on Wildlife**

Venue	Witness Tone			
(Committees)	Private Development and Trade Tone	Environmental Protection Tone	Neutral or Uncodeable Tone	Totals (N)
Private Development and Trade Tone	42.5% (512)	45.5% (548)	12% (145)	100% (1205)
Environmental Protection Tone	25.7% (726)	63.8% (1875)	11.5% (337)	100% (2938)
Neutral or Uncodeable Tone	7.5% (23)	83.8% (258)	8.8% (27)	100% (308)
Totals	28.3% (1261)	60.2% (2681)	11.4% (509)	100% (4451)

Gamma = 0.29, Kendall's Tau-b = 0.15, Chi-Square = 227 (p < 0.00001)

by others. An ordinary least squares regression, used to test how much Congressional attention is paid to wildlife, explains the growth of venues claiming wildlife jurisdiction. Equation 5.1 shows the ordinary least squares equation used in this analysis. Congressional attention, measured as the number of hearings on wildlife per year (independent variable), results in increases to the number of available wetlands venues as measured by the number of subcommittees (dependent variable).

Equation 5.1 Congressional Attention and the Role of Committees

$$Y_t = \beta_1 X_t + \beta_2 Y_{t-1} + e_t$$

Y_t = Number of Hearings conducted each year on wildlife policy. (dependent variable)

X_t = Congressional Committees holding Hearings on wildlife policy each year.

Y_{t-1} = Number of Hearings conducted previous year on wildlife policy. (lag variable)

e_t = residual unexplained by model.

The regression results in Table 5.4 indicate that as Congressional attention increases, Congressional venues proliferate to accommodate venue-shoppers and committees seeking to expand their jurisdiction (adjusted R Square = .76; Beta = .69; p < .0001). In fact, the results show that Congressional attention has a large impact on the expansion of venues. A nice triangulation of results is demonstrated in Table 5.1 when the Chi-Square tests and regression analysis are taken into consideration. These results point to a valence issue of environmental protection for wildlife issues with some opposition coming from the private development and trade interests who are seeking to expand the definition in Congress.

Wildlife as a Valence Issue Model

Two important conclusions are drawn from this study on wildlife policymaking in Congress. First, the original and persistent definition of wildlife in Congress has assisted in the recovery and prevention of loss of wildlife. While no major redefinition process has occurred, there is an underlying competing definition to wildlife issues that focuses on private development and trade conflicts. This alternative definition, while being represented in Congress, does not represent a powerful influence on wildlife policies. In addition, there has been minimal impact on wildlife as a natural resource from this competing

**Table 5.4 Relationship between Congressional Venues and Hearings
in Great Lakes Polciy from 1789 to 1999**

Variable	B	Standard Error	Beta Coefficient	Significance
Intercept	0.45	0.14		< 0.0001
Wildlife Hearings	0.391	0.03	0.82	< 0.0001
F Statistic				< 0.0001
Adjusted R-Square =	0.77			
Durbin Watson* =	1.79			
N =	112			

*Indicates no autocorrelation problem. The dependent variable is annual number of subcommittees holding hearings on Great Lakes.

definition. Conflicts have been infrequent and contained to certain local incidents that have not altered the basic definition of the wildlife policies. Second, as a valence issue, wildlife policies may continue to enjoy the safety of being somewhat monotonic as compared to the disruption that results from dominance issues redefinitions. However, if the private development and trade alternative definitions gain momentum in the strategy of expanding its jurisdiction by creating more venues in Congress to assist in the venue-shopping process, this could lead to a change in wildlife policy. It is not impossible for policies to switch issue definition models. For instance, the Great Lakes policies initially displayed a typical dominance issue model until Era III when they took the path of a bounded model. The advantages of the bounded model are that interests are not completely replaced but incorporated into the redefinition of the issue. The Merchant Marines and Fisheries Committee in the Great Lakes policy exhibited this adapting ability to have multiple tones sharing the agenda. It is this same committee that dominates wildlife policy, yet the tone remains monotonic. Again, while wildlife policies enjoy the secure place of being a valence issue (which does make it difficult to mobilize against due to powerful symbols) it is not inconceivable that in the future wildlife policies could undergo a redefinition process in Congress under this committee. From the data, a redefinition seems unlikely since the definition of wildlife policies as one of an environmental protection tone has lasted since its original conception as a public trust. However, there is nothing that precludes significant events from occurring to favor a redefinition process in Congress. No doubt history has shown that it is certainly more difficult to mount opposition strong enough to redefine a valence issue. Nonetheless, the next several decades may show a shift from the stability of valency to one of disruption from a redefinition process.

CHAPTER 6

Effectiveness of Using Issue Definition Models for Understanding Agenda Setting in Congress

There are several conclusions that can be made regarding how issue definitions can be used to understand agenda setting in Congress. The link between issue definition and agenda setting in Congress is shown to be critical to implementation of public policies, particularly environmental policies. It is expected that the same linkage has just as critical an impact on other policies as well. One reason natural resource cases of wetlands, the Great Lakes, and wildlife were chosen is because there are visual impacts which are easily viewed and directly linked to how the natural resource was defined in Congress. Four major conclusions of using issue definition models for understanding agenda setting in Congress are presented in this chapter. These conclusions are discussed in the next section. First, a definition of an issue can be identified through the tone of the hearing and the participants. Second, redefinition tools of legislative and nonlegislative hearings may not be the most useful in understanding how tone can change for an issue. Jurisdictional expansion in Congress through adaptation of committees' definition of issues, as well as introduction of new committees are more effective in describing how issues are redefined than legislative and nonlegislative hearings. Venue-shopping does occur in all the issue definition models regardless of tone. Third, issues have the ability to change models over time which may mean different outcomes for policies. Finally, the classical question of who participates and how democratic that participation is in Congress has a surprising outcome when witnesses are tracked for the life cycle of an issue. Each one of these conclusions is further detailed below along with some recommendations.

Understanding Issue Definition using Tone

Clearly from the results of Chapters Three through Five, definition of an issue can be tracked over time in Congress by identifying the tone of hearings, committees, and witnesses who testify. The tone of an issue is classified into a dichotomous variable by first identifying the baseline or original tone of an issue. This allows the framework for looking for redefinitions or change to the tone of an issue which usually signals changes that come from uses of symbols, rhetoric, and policy implementation. The three issue definitions models provide an effective typology for predicting agenda setting behavior of participants in the policymaking process. The characteristics of these models are premised on tone, symbols, rhetoric, and policy implementation. The patterns described for each model based on indicators of hearing type, in addition to number of hearings, committees, and witnesses provide scholars of congressional policy, agenda setting, and public policy better insight toward understanding how policies change over time. From the results of applying the issue definition models to wetlands, the Great Lakes, and wildlife, the linkages between public policy and agenda setting are better understood and explained.

The Redefinition Tools of Legislative and Nonlegislative Hearings

Using the redefinition tools of nonlegislative hearings to predict periods of issue redefinition in each model failed. What is interesting is that in not one case did nonlegislative hearings prove helpful for breaking up the status quo or baseline definition of a policy. Original policies were not altered by committees who may be seeking to expand jurisdiction using nonlegislative hearings to provide competing definition of an issue. In fact, in the Great Lakes case, just the opposite is true. Legislative hearings maintained the majority of hearing type used throughout the entire life cycle of policymaking for the Great Lakes. Likewise, wetlands policy had the largest majority of nonlegislative hearings during a period of dominance by agricultural interests not a period of redefinition. Therefore, using hearing type did not help in predicting the different dominant tone of an issue. Perhaps this is something unique to environmental policies; nonetheless, future research using nonlegislative hearings as a means to expand jurisdictions or change issue definition should be subject to skepticism.

Jurisdictional expansion in Congress through adaptation of committees' definition of issues as well as introduction of new committees was more effective in describing how issues are redefined than legislative and nonlegislative hearings. While one method for expansion is the use of nonlegislative hearings, Congress did not rely on this tool in the cases of wetlands, the Great Lakes, or wildlife policymaking. Interestingly, based on the statistical results, venue-shopping does occur in all the issue definition models regardless of tone. More committees conduct hearings to assist the process of issue redefinition over time. The ability of committees to adapt and incorporate a new definition and tone of an issue was displayed by the Merchant Marines and Fisheries Committee. This venue displayed the benefits of being able to incorporate the new definition of an issue by making it part of its jurisdiction. The outcome from a challenging tone of an issue being folded into the existing venue had positive benefits for the Great Lakes. This phenomenon which takes place in the bounded issue model hoods great benefits for policy outcomes.

Changing Issue Definition Models over Time

One unexpected conclusion from taking a longitudinal approach to issue definition was showing how issues on one issued model path are not necessarily locked into that model for the life span of the issue. The Great Lakes issue definition process was closely following the dominance issue model pattern of wetlands. This path could have lead to the high levels of conflict, chaotic policy implementation, and negative effects from a dominance model similar to the wetlands. However, the issue was transformed into taking a bounded issue path, which has proven to be beneficial to restoring the Great Lakes and reversing a significant amount of damage in a short period of time. This is not the case for wetlands. While today, there is no net loss of wetlands, it is not clear if wetlands are better protected, yet alone, being restored to previous numbers across the country.

Who Participates and How Democratic
is Participation in Congress?

One very surprising outcome of all three models of issue definition is that no one type of witness or single witness dominates the policy. When witnesses were tracked over time, not a single witness received more than six percent of participation frequency for the life cycle of an issue. Some types of witnesses may have a larger percentage in a particular era, but none testified before Congress in large percentages over time. This is surprising because often in the media, interest groups with tones of environment, private sector, or agricultural tones are portrayed as dominating a particular policy. Even the traditional iron triangle theories of public policy portray a sector of interests as being the main participant in policymaking in Congress. However, Congress does have opportunities for a variety of witnesses to testify. This leads one to conclude that participation by witnesses is more democratic than previous theories and contemporary wisdom lead us to believe. While congressional committees tend to hear from witnesses with a policy agenda much like its own, there are opportunities for venue-shopping where committees are created, adapted, and competed for explaining jurisdictions which again create opportunity for more democratic participation.

Where Congress appears less democratic is in who it calls as witnesses. A vast majority of witnesses called by Congress to participate in hearings are males. Based on the data collected for wetlands, the Great Lakes, and wildlife, most participants contributing to the issue definition process are male. This is not surprising since committees tend to rely on witnesses of the same tone of an issue, and on witnesses with the same gender. This also may change over time.

Perhaps unique to environmental policies, there was a tendency for the federal government to become more relied on as a witness when the tone of the issue became protection of the natural resource, except for the case of wetlands policy. What was supportive of Congress remaining open to democratic participation based on variety of witnesses was that private sector interests were never a major participant in any of the cases.

Achieving and Ecosystem and Sustainable Development Approach for Management of Natural Resources Through Issue Definition

One of the most important findings in this book is the management of natural resources. The link between agenda setting and issue definition has a significant role to understand how to manage a natural resource. How the natural resource is defined by Congress and participants in congressional hearings has shown in the cases of wetlands, the Great Lakes and wildlife to have dramatic impact on the sustainability of the resource. A very practical application of this approach of using issue definition models is to move an issue to a bounded issue model type. While there are benefits to the valence issue model, clearly if sustainability and an ecosystem approach is desired for management of the resource, a bounded issue model is the optimal path. The goal of international and national organizations interested in natural resource policy has been to seek policies that seek the most efficient use of a resource which include not exhausting it through overconsumption. A dominance issue model does not address issues of overconsumption of a natural resource because multiple definitions of an issue are not feasible. A valence issue model could provide maximum protection of the natural resource but this does not achieve sustainability or use of the resource. Clearly, a valence issue model assists a natural resource like wildlife since it can prevent nonreversible outcomes such as extinction and bring species back from being threatened with extinction. Any resource perhaps on the path of being completely and nonreversibly exhausted could benefit from a valence issue model path where human extraction of the resource can be extremely limited. However, more natural resources require a balance of human consumption and continued health that has been termed a sustainable pattern of use of ecosystems. Policymakers in Congress, implementing agencies, and other participants in the agenda setting arena could learn of the bounded issue model from the benefits to the Great Lakes. Many of these policy participants are seeking paths to achieving ecosystem management not because the science is not available as tools but the agenda setting process is unclear. Using issue definition models, and understanding tone could provide policymakers with the answers they have sought to achieve sustainable policies for natural resources.

Bibliography

Aldrich, John H. (1995). *Why Parties?* Chicago, IL: University of Chicago Press.

Allance, David R. and Thorp, Steve A. (1995). Challenging the Great Lakes Economy: Economic and Environmental Linkages, Background Paper for the State of the Lakes Economic System Conference. Dearborn, Michigan: Environment Canada and U.S. U.S. EPA, EPA 905-R-95–017.

Bardach, E. (1977). The Implementation Game. In *Public Policy: The Essential Readings*, edited by S. Z. Theodoulou and M. A. Cahn. Englewood Cliffs, NJ: Prentice Hall Press.

Baumgartner, Frank and Jones, Bryan. (1993). *Agendas and Instability in American Politics*. Chicago, IL: University of Chicago Press.

Baumgartner, Frank, Jones, Bryan, and MacLeod, Micheal. (1998). Lesson from the Trenches: Ensuring Quality, Reliability, and Usability in the Creation of a New Data Source. *The Political Methodologist* 8: 1–10.

Bean, Michael J. (1983). *The Evolution of National Wildlife Law: Revised & Expanded Edition*. New York: Praeger Publishers.

Bernstein, Robert. (1989). *Elections, Representation, and Congressional Voting Behavior.* Englewood Cliffs, NJ: Prentice Hall Press.

Berry, Jeffrey. (1989). "Subgovernments, Issue Networks, and Political Conflict." In *Remaking American Politics*, edited by Richard, Harris, and Sidney, Milks. Boulder: Westview.

Bosso, Christopher. (1987). *Pesticides and Politics: The Life Cycle of a Public Issue*. Pittsburgh: University of Pittsburgh.

Botts, Lee and Muldoon, Paul. (1996). *The Great Lakes Water quality Agreement: Its Past Successes and Uncertain Future*. Hanover, New Hampshire: Dartmouth College.

Bowles, Marlin L. and Whelan, Christopher J. (1994). *Restoration of Endangered Species: Conceptual Issues, Planning, and Implementation*. Cambridge, Britain: Cambridge University Press.

Boylan, Karen D. and MacLean, Donald R. (1997). Linking Species Loss with Wetlands Loss. *National Wetlands News Letter* 19: 6.

Bryner, Gary. (1987). *Blue Skies, Green Politics: The Clean Air Act of 1990*. Washington, D.C.: Congressional Quarterly Press.

Campbell, Faith. (1991). "The Appropriations History." In *Balancing on the Brink of Extinction: The Endangered Species Act and Lessons for the Future*, edited by Kathryn A. Kohm. Washington, D.C.: Island press.

Campbell, Cathryn. (1983). Federal Protection of Endangered Species: A Policy of Overkill? *UCLA Journal of Environmental Law and Policy* 3 (2): 247–275.

Cater, Douglas. (1964). *Power in Washington*. New York: Random House.

Cheever, Federico. (1997). The Road to Recover: A New Way of Thinking About the Endangered Species Act. *Land use & Environmental Law* 28: 549–627.

Cobb, Roger W., and Elder, Charles D. (1983). *Participation in American Politics: The Dynamics of Agenda-Building*. Baltimore, MD: Johns Hopkins University Press.

Coggins, George C. (1973). Conserving Wildlife Resources: An Overview of the Endangered Species Act of 1973. *North Dakota Law Review* 51: 315–337.

Coggins, George C. (1991). "Snail Darters and Pork Barrels Revisited: Reflections on Endangered Species and land Use in America." In *Balancing on the Brink of Extinction: The Endangered Species Act and lessons for the Future*, edited by Kathryn A. Kohm. Washington, D.C.: Island press.

Colborn, Theodora, and Davidson, A., Green, S. N., Hodge, R. A., Jackson, C. I., and Liroff, R. A. (1990). The Great Lakes, *Great Legacy?* Washington D.C.: Conservation Foundation, and Ottawa, Canada: Institution on Public Policy.

Cone, Marla. (1995). Endangered Species Act is Sound Science, Panel Finds. *Los Angeles Times*, May 25, 1995.

Davidson, Roger. (1977). "Breaking up those Cozy Triangles: An impossible dream?" In *Legislative Reform and Public Policy*, edited by Susan Welch, and Guy Peters. New York: Praeger.

Dahl, Robert. (1961). *Who Governs?*. New Haven: Yale University Press.

Dahl, Thomas E. (1990). *Wetlands Losses in the United States 1780s to 1980s*. Washington D.C: U.S. FWS.

Dahl, Thomas E. (1991). *Wetlands Status and Trends in the Conterminous United States mid-1970s to mid-1980s*. Washington, D.C.: U.S.: FWS.

Dodd, Lawrence and Oppenheimer, Bruce. (1999). *Congress Reconsidered*. Washington, D.C.: CQ Press.

Dennison, Mark, S., and Berry, James F. (1993). *Wetlands—Guide to Science, Law and Technology*. Park Ridge, NJ: Noyes Publications.

Dingell, John D. (1991). "The Endangered Species Act: Legislative Perspectives on a Living Law." In *Balancing on the Brink of Extinction: The Endangered Species Act and lessons for the Future*, edited by Kathryn A. Kohm. Washington, D.C.: Island press.

Downs, Anthony. (1972). Up and Down with Ecology: The Issue Attention Cycle. *Public Interest* 28: 38–50.

Dunlap, Riley and Mertig, Angela. (1992). *American Environmentalism: The U.S. Environmental Movement 1970–1990*. Washington D.C.: Taylor and Francis.

Dye, Thomas and Zeigler, Harmon (1999). *The Irony of Democracy*. Fort Worth, TX: Harcourt Brace College Publishers.

Edelman, Murray. (1974). *Symbols as Symbolic Action*. Chicago: Markham Press.

Ehrlich, Paul and Ehrlich, Anne. (1981). *Extinction*. New York: Randam House.

Fenno, Richard. (1973). *Congressman in Committees*. Boston: Little Brown.

Fiorinia, Morris. (1989). *Congress: Keystone of the Washington Establishment*. New Haven: Yale University Press.

Fischhoff, Baruch, Derby, Stephen, and Lichtenstein, Sarah. (1984). *Acceptable Risk*. Cambridge: Cambridge University Press.

Fitzgerald, Sarah. Wildlife Trade: Whose Business is it. Washington D.C.: World Wildlife Federation.

Freeman, J. Leiper. (1965). *The Political Process: Executive Bureau-Legislative Committee Relations*. New York: Random House.

Fredman, Peter and Boman, Mattias. (1996). Endangered Species and
     Optimal Environmental Policy. *Journal of Environmental Management*
     47: 381–389.

Gleivck, Peter H. (ed.) (1993). *Water in Crisis: A Guide to the World's
     Fresh Water Resources.* New York: Oxford University Press.

Greenwalt, Lynn A. (1991). "The Power and Potential of the Act." In
     *Balancing on the Brink of Extinction: The Endangered Species Act and
     Lessons for the Future*, edited by Kathryn A. Kohm. Washington, D.C.:
     Island Press.

The Great Lakes Water Quality Agreement, signed April 14, 1972.
     International Commission, Ottawa and Washington, D.C.

The Great Lakes Water Quality Agreement. (May 12, 2000). Available:
     http://www.cciw.ca/glwqa/history-backgrounder-e.ht*ml*.

Griffith, Ernest S. (1939). *The Impasse of Democracy.* New York: Harrison-
     Hilton Books.

Hall, Richard. (1996). *Participation in Congress.* New Haven, Conn. Yale
     University Press.

Hall, Richard and Evans, Lawrence C. (1990). The Power of
     Subcommittees. *Journal of Politics* 52: 335–354.

Heclo, Hugh. (1978). "Issue Networks and the Executive Establishment." In
     The New American Political System, edited by Anthony King.
     Washington, D.C.: American Enterprise Institute.

Hinckley, Barbara. (1975). Policy Content, Committee Membership, and
     Behavior. *Journal of Political Science* 19: 543–548.

Interagency Working Group on Federal Wetlands Policy. (1993*).
     Agriculture and Wetlands: The Swampbuster Program.* Washington,
     D.C.: Congressional Quarterly Press.

Heclo, Hugh. (1979). "Issue Networks in the Executive Establishment." In
     The New American Political System, edited by Anthony King.
     Washington, D.C.: American Enterprise Institute.

Hill, Kevin D. (1993). The Endangered Species Act: What Do We Mean By
     Species? *Environmental Affairs* 20: 239–264.

John, Dewitt. (1994). *Civic Environmentalism: Alternative to regulation in
     the states.* Washington, D.C.: Congressional Quarterly Press.

Jones, Bryan, Frank, Baumgartner, and Talbert, Jeffrey. (1993). The
     Destruction of Issue Monopolies in Congress. *American Political
     Science Review* 87: 657–671.

King, David C. (1994). The Nature of Congressional Committee Jurisdictions. *American Political Science Review* 88: 1, pp. 48–62.

King, David C. (1997). *Turf Wars: How Congressional Committees Claim Jurisdiction.* Chicago, IL: University of Chicago Press.

Kingdon, John W. (1984). *Agendas, Alternatives, and Public Policies.* Boston: Little, Brown.

Kirp, David. (1982). Professionalization as a Choice. *World Politics* 34: 137–174.

Knickerbocker, Brad. (1995). Endangered Species Act Faces its Own Dangers. *Christian Science Monitor*, (March 3, 1995).

Kohm, Kathryn A. (1991). "The Act's History and Framework." In *Balancing on the Brink of Extinction: The Endangered Species Act and lessons for the Future*, edited by Kathryn A. Kohm. Washington, D.C.: Island press.

Krehbiel, Keith. (1993). Where's the Party? *British Journal of Political Science* 23: 235–66.

Kubasek, Nancy K. and Browne, Neil M. (1994). The Endangered Species Act: An Evaluation of Alternative Approaches. *Dickinson Journal of Environmental Law and Policy* 3 (2): 1–18.

Kukoy, S. and Canter, L. (1995). "Mitigation Banking as a Tool for Improving Wetlands Preservation via Section 404 of the Clean Water Act." *The Environmental Professional* 17: 301–308.

Kessler, Jon A. (1990). *Our National Wetlands Heritage.* Washington, D.C.: Environmental Law Institute.

Lents, Michael. (1994). President Clinton and Wetlands Regulations: Boon or Bane to the Environment? *Temple Environmental Law and Technology Journal* 13: 81–103.

Lester, James and Lombard, E. (1990). The Comparative Analysis of State Environmental Policy. *Natural Resources Journal* 30: 301–19.

Lowi, Theodore J. (1964). American Business, Public Policy, Case Studies, and Political Theory. *World Politics* 16: 677–693.

Lowi, Theodore. (1972). Four Systems of Policy, Politics, and Choice. *Public Administration Review* 32: 298–210.

March, J., Cohen, J. and Olsen, J. (1972). Garbage Can Model of Organizational Choice. *Administrative Science Quarterly* 17 (1): 1–25.

Mayhew, David. (1986). *Congress the Electoral Connections.* New Haven: Yale University Press.

McConnell, Grant. (1967). *Private Power and American Democracy*. New York: Alfred Knopf.

Mitch, William J. and Gosslink, James G. (1993). *Wetlands*. New York: Van No strand Reinhold.

Majone, Giordano and Wildavsky, Aaron. (1984). "Implementation As Evolution." In *Public Policy: The Essential Readings,* edited by S. Z. Theodoulou, and M. A. Cahn. Englewood Cliffs, NJ: Prentice-Hall.

Mann, Charles C. and Plummer, Mark L. (1995). *Noah's Choice: The Future of Endangered Species*. New York: Alfred A. Knopf, Inc.

Mayhew, David. (1974). *Congress: The Electoral Connection*. New Haven, Conn: Yale University Press.

Mazmanian, Daniel and Sabatier, Paul. (1983). *Implementation and Public Policy*. Glenview: Scott, Foresman, and Company.

Meyers, Gary D. (1991). Old-Growth Forests, The Owl, and Yew: Environmental Ethics Versus Traditional Dispute Resolution Under the Endangered Species Act and Other Public Lands and Resources Laws. *Environmental Affairs* 18 (4): 623–668.

Morone, Joseph and Woodhouse, Edward. (1989). *The Demise of Nuclear Energy?* New Haven: Yale University Press.

National Audubon Society. (1996). *Small and Farmed Wetlands: Oasis for Wildlife*. Washington, D.C.: PSA, Inc.

National Research Council. (1995). *Wetlands: Characteristics and Boundaries*. Washington, D.C.: National Academy Press.

Nelkin, Dorothy. (1971). *Nuclear Power and Its Critics*. Ithaca, NY: Cornell University Press.

Nelkin, Dorothy and Fallows, Susan. (1978). The Evolution of the Nuclear Debate. *Annual Review of Energy* 3: 275–312.

Phelps, Martha F. (1997). Candidate Conservation Agreements Under the Endangered Species Act. *Boston College Environmental Affairs Law Review* 25 (1): 175–213.

Palmer, William D. (1975). Endangered Species Protection: A History of Congressional Action. *Environmental Affairs* 4: 255–293.

Peltzman, Samuel. (1984). Constituent Interest and Congressional Voting. *Journal of Law and Economics,* 27: 181–210.

Pressman, Jeffrey and Wildavsky, Aaron. (1973). *Implementation*. Berkeley: University of California.

Price, David. (1978). Policymaking in Congressional Committees. *American Political Science Review* 72: 548–574.

Rabe, Barry. (1996). An Empirical Examination of Innovations in Integrated Environmental Management: The Case of the Great Lakes. *Public Administration Review* 56: 4, 372–381.

Redford, Emmette S. (1969). *Democracy in the Administrative State*. New York: Oxford University Press.

Revised The Great Lakes Water Quality Agreement of 1978, with Annexes and Terms of Reference, Signed November 22, 1978. International Joint Commission, Ottawa and Washington D.C.

Rhode, David W. (1991). *Parties and Leaders in the Postreform House*. Chicago, IL: University of Chicago.

Rhodes, Richard. (1987). *The Making of the Atomic Bomb*. New York: Touchstone Books.

Rohlf, Daniel. (1991). Six Biological Reasons Why the Endangered Species Act Doesn't Work: And What to Do About it. *Consevration Biology* 5 (3): 273–290.

Riker, William H. (1986). *The Art of Political Manipulation*. New Haven, CT: Yale University Press.

Ringquist, Evan. (1993). *Environment Protection at the State Level: Politics and Progress in Controlling Pollution*. Armonk, NY: ME Sharpe.

Ripley, Randall and Franklin, Grace. (1987). *Congress, the Bureaucracy and Public Policy*. 4th ed. Chicago: Dorsey Press.

Rochefort, D. and Cobb, Roger. (1994). *The Politics Problem Definition: Shaping the Policy Agenda*. Lawrence, Kansas: University Press of Kansas.

Rogers, E. and Dearing, John. (1988). Agenda-Setting Research In James Anderson, ed., *Communication Yearbook* 11 (555–594). Newbury Park, CA: Sage.

Rogers, E., Dearing, John and Bregman, Daniel. (1993). The Anatomy of Agenda-Setting Research. *Journal of Communication 43* (2): 68–84.

Rohlf, Daniel J. (1989). *Endangered Species Act*. Stanford, CA: Stanford Environmental Law Society.

Rolston III, Holmes. (1991). "Life in Jeopardy on Private Property." In *Balancing on the Brink of Extinction: The Endangered Species Act and lessons for the Future*, edited by Kathryn A. Kohm. Washington, D.C.: Island press.

Sax, Joseph L. (1997). New Departures in the Legal Protection of Biological Diversity: Implementing the Endangered Species Act. *Environmental Policy and Law* 27(4): 347–350.

Sabatier, Paul and Mazmanian, Daniel. (1980). "A Conceptual Framework of the Implementation Process." In *Public Policy: The Essential Readings*, edited by S. Z. Theodoulou, and M. A. Cahn. Englewood Cliffs, NJ: Prentice-Hall.

Salwasser, Hal. (1991). "In Search of an Ecosystem Approach to Endangered Species Conservation." In *Balancing on the Brink of Extinction: The Endangered Species Act and lessons for the Future*, edited by Kathryn A. Kohm. Washington, D.C.: Island Press.

Salsburg, Robert. (1984). Interest Representation. *American Political Science Review* 78: 64–76.

Schattschneider, E. E. (1960). *The Semi-Sovereign People*. New York: Holt, Rinehart, Winston.

Shepsle, Kenneth and Weingast, Barry. (1987). The Institutional Foundations of Committee Power. *American Political Science Review* 81: 1 86–104.

Stalcip, Brenda, (eds). (1996). *Endangered Species Opposing Viewpoints*. San Diego, CA: Green Haven Press Inc.

Scher, Seymour. (1960). Congressional Committee Members as Independent Agency Overseers. *American Political Science Review*, 54: 911–920.

Silverstein, Jonathan. (1994). Taking Wetlands to the Bank. *Boston College Environmental Affairs Law Review* 22: 29–161.

Sohn, David and Cohen, Madeline. (1996). From Smokestacks to Species: Extending the Tradable Permit Approach from Air Pollution to Habitat Conservation. *Stanford Environmental Law Society* 15 (2): 406–450.

Stone, Deborah. (1988). *Policy Paradox and Political Reason*. Glenview, IL: Scott, Foresman.

Talbert, Jeffrey and Baumgartner, Frank R. (1995). Nonlegislative Hearings and Policy Change in Congress. *American Journal of Political Science* 39: 383–405.

Tarlock, Dan A. 1991. Western Water Rights and the Endangered Species Act. In *Balancing on the Brink of Extinction: The Endangered Species Act and lessons for the Future*, edited by Kathryn A. Kohm. Washington, D.C.: Island Press.

Tzoumis, Kelly. (1998). Wetland Policymaking in the U.S. Congress From 1789 to 1995. *Wetland* 18 (3): 447–459.

Weart, Spencer. (1988). *Nuclear Fear: A History of Images*. Cambridge, MA: Harvard University Press.

Vig, Norman and Kraft, Michael. (1996). *Environmental Policy in the 1990s: Reform or Reaction*. Washington, D.C.: Congressional Quarterly Press.

Turini, A. (1991). "Swampbuster: A Report from the Front" *Indiana Law Review* 24, 1507–1524.

U.S. Department of Interior. (1994). "The Impact of Federal Programs on Wetlands: A Report to Congress." Washington. D.C.: U.S. Department of Interior.

U.S. EPA. (1995). The Great Lakes: An Environmental Atlas and Resource Book. 3 edition. Chicago, IL: EPA #905-B-95–001.

Vig, Norman and Kraft, Michael. (1996). *Environmental Policy in the 1990s: Reform or Reaction*. Washington, D.C.: Congressional Quarterly Press.

White House. (1993a). A Summary of Protecting America's Wetlands: A Fair, Flexible, and Effective Approach. Washington, D.C.: White House.

White House. (1993b). New Federal Wetlands Policy Offers Fair, Flexible Approach Ends Agency Infighting and Gridlock with Strong Agreement. Washington, D.C.: Office of Environmental Policy.

White House. (1995). *The Clinton Administration Wetlands Plan: An Update*. Washington, D.C.: White House.

Wood, B. Dan and Waterman, Richard. (1991). The Dynamics of Political Control of the Bureaucracy. *American Political science Review* 85: 3: 801–828.

Yaffee, Steven L. (1991). "Avoiding Endangered Species: Development Conflicts Through Interagency Consultation." In *Balancing on the Brink of Extinction: The Endangered Species Act and Lessons for the Future*, edited by Kathryn A. Kohm. Washington, D.C.: Island Press.

Yaffee, Steven L. (1994). *The Wisdom of the Spotted Owl: Policy Issues for a New Century*. Washington D.C.: Island Press.

Zucciputti, John A. (1995). A Native Returns: The Endangered Species Act and Wolf Reintroduction to the Northern Rocky Mountains. *Columbia Journal of Environmental Law* 20 (2): 329–360.

Index